WINNING THE ARCH WAY

THE STORY OF ARCH ALUMINUM & GLASS

CASINO
GALLERIA
CASINO

WINNING THE ARCH WAY

THE STORY OF ARCH ALUMINUM & GLASS

JEFFREY L. RODENGEN

Edited by Anne Forsyth and Ann Gossy
Design and layout by Ryan Milewicz

PHOTO COURTESY OF IDS PHOTOGRAPHY.

WRITE STUFF

Write Stuff Enterprises, Inc.
1001 South Andrews Avenue
Fort Lauderdale, FL 33316
1-800-900-Book (1-800-900-2665)
(954) 462-6657
www.writestuffbooks.com

Publisher's Cataloging in Publication
(Prepared by The Donohue Group, Inc.)

Rodengen, Jeffrey L.
Winning the Arch Way : the story of Arch Aluminum & Glass / Jeffrey L. Rodengen ; edited by Anne Forsyth and Ann Gossy ; design and layout by Ryan Milewicz ; [foreword by Leon Silverstein].

p. : ill. ; cm.

Includes bibliographical references and index.
Issued also as CD-ROM.
ISBN-13: 978-1-932022-23-0
ISBN-10: 1-932022-23-6

1. Arch Aluminum and Glass (Firm)—History. 2. Aluminum industry and trade—United States—History. 3. Architectural glass—United States—History. I. Forsyth, Anne (Anne V.), 1983- II. Gossy, Ann. III. Milewicz, Ryan. IV. Silverstein, Leon. V. Title. VI. Title: Arch way

HD9539.A64 A734 2008
338.7/673722/0973 2008938233

Completely produced in the
United States of America
10 9 8 7 6 5 4 3 2 1

Also by Jeffrey L. Rodengen

The Legend of Chris-Craft

IRON FIST: The Lives of Carl Kiekhaefer

Evinrude-Johnson and The Legend of OMC

Serving the Silent Service: The Legend of Electric Boat

The Legend of Dr Pepper/Seven-Up

The Legend of Honeywell

The Legend of Briggs & Stratton

The Legend of Ingersoll-Rand

The Legend of Stanley: 150 Years of The Stanley Works

The MicroAge Way

The Legend of Halliburton

The Legend of York International

The Legend of Nucor Corporation

The Legend of Goodyear: The First 100 Years

The Legend of AMP

The Legend of Cessna

The Legend of VF Corporation

The Spirit of AMD

The Legend of Rowan

New Horizons: The Story of Ashland Inc.

The History of American Standard

The Legend of Mercury Marine

The Legend of Federal-Mogul

Against the Odds: Inter-Tel—The First 30 Years

The Legend of Pfizer

State of the Heart: The Practical Guide to Your Heart and Heart Surgery
with Larry W. Stephenson, M.D.

The Legend of Worthington Industries

The Legend of IBP

The Legend of Trinity Industries, Inc.

The Legend of Cornelius Vanderbilt Whitney

The Legend of Amdahl

The Legend of Litton Industries

The Legend of Gulfstream

The Legend of Bertram
with David A. Patten

The Legend of Ritchie Bros. Auctioneers

The Legend of ALLTEL
with David A. Patten

The Yes, you can of Invacare Corporation
with Anthony L. Wall

The Ship in the Balloon: The Story of Boston Scientific and the Development of Less-Invasive Medicine

The Legend of Day & Zimmermann

The Legend of Noble Drilling

Fifty Years of Innovation: Kulicke & Soffa

Biomet—From Warsaw to the World
with Richard F. Hubbard

NRA: An American Legend

The Heritage and Values of RPM, Inc.

The Marmon Group: The First Fifty Years

The Legend of Grainger

The Legend of The Titan Corporation
with Richard F. Hubbard

The Legend of Discount Tire Co.
with Richard F. Hubbard

The Legend of Polaris
with Richard F. Hubbard

The Legend of La-Z-Boy
with Richard F. Hubbard

The Legend of McCarthy
with Richard F. Hubbard

Intervoice: Twenty Years of Innovation
with Richard F. Hubbard

Jefferson-Pilot Financial: A Century of Excellence
with Richard F. Hubbard

The Legend of HCA

The Legend of Werner Enterprises
with Richard F. Hubbard

The History of J. F. Shea Co.
with Richard F. Hubbard

True to Our Vision
with Richard F. Hubbard

The Legend of Albert Trostel & Sons
with Richard F. Hubbard

The Legend of Sovereign Bancorp
with Richard F. Hubbard

Innovation is the Best Medicine: The extraordinary story of Datascope
with Richard F. Hubbard

The Legend of Guardian Industries

The Legend of Universal Forest Products

Changing the World: Polytechnic University—The First 150 Years

Nothing is Impossible: The Legend of Joe Hardy and 84 Lumber

In it for the Long Haul: The Story of CRST

The Story of Parsons Corporation

Cerner: From Vision to Value

New Horizons: The Story of Federated Investors

Office Depot: Taking Care of Business—The First 20 Years

The Legend of General Parts: Proudly Serving a World in Motion

Bard: Power of the Past, Force of the Future

Innovation & Integrity: The Story of Hub Group

Amica: A Century of Service 1907–2007

A Passion for Service: The Story of ARAMARK

The Legend of Con-way: A History of Service, Reliability, Innovation, and Growth

Commanding the Waterways: The Story of Sea Ray

Past, Present & Futures: Chicago Mercantile Exchange

The Legend of Leggett & Platt

Table of Contents

Foreword by Leon Silverstein vii

Acknowledgments viii

Chapter I Stepping Toward the Future 10

Chapter II A Growing Enterprise 22

Chapter III The Chromalloy Years 34

Chapter IV Move Over, Miami 42

Chapter V Innovation and Growth 56

Chapter VI On the Edge of the Precipice 74

Chapter VII Innovation and Expansion 88

Appendix Arch Aluminum & Glass' Signature Projects 114

Notes to Sources 124

Index 138

FOREWORD

BY

LEON SILVERSTEIN

CHIEF EXECUTIVE OFFICER, ARCH ALUMINUM & GLASS

MY FATHER, BOBBY SILVERSTEIN, started Arch Aluminum & Glass from scratch at the age of 43. He abandoned a lucrative corporate managerial position, uprooted his family, and rolled the dice by launching his own business. This was true to his character, as he has essentially gambled for his entire life. A very competitive person, my father fought to make a living, and he fought to win.

At Arch, we believe winning is about building customer loyalty and promoting high employee morale. Winning is about creating opportunities for employees—then watching them grow—and watching them win. Winning is selling service, not price. Winning is about growing the business, because if your business doesn't grow, it will surely die. Each of these philosophies originated with my father. We have followed his lead and adopted his philosophy and his culture. Everything we do at Arch is an extension of my father and his vision. His values served as the solid foundation from which we built a successful business.

Through his quiet leadership and hard work, my father developed a winning business. His greatest strength has always been how well he relates to people. Employees love working for him. He's more comfortable with the workers on the front lines than he is with the executive types. He takes great pleasure in looking back at the guys who were on the floor 10, 15, 20 years ago, and seeing how they advanced to their positions today. Although he may be the least visible player from the competitors' perspective, he is the most visible from an employee perspective.

I share my father's competitive spirit. As a child, I remember watching him play basketball. He was a nut case on the court. He didn't care if the other player was 6' 6" and 280 lbs. He would challenge anybody, but he always played fair. That competitive nature and fair-mindedness helped form the basis for our business. My father's focus on sports and competition has played an integral role in molding our business philosophy.

People constantly ask me, "What is Arch's end game? What is the end game for the family?" I always answer the same way. "There is no end game. We're all workers first, and we're at the top of our game. Why would we ever sell our business when we love it and it's our legacy?" My son, Jacob, understands that the business is our family legacy. At his bar mitzvah, he told everyone that someday he will be working in the family business.

A favorite expression of my father's is, "There's no Monday morning quarterback. You don't look back." We live by his motto, and we don't dwell on the past. However, the past can serve as a bridge to the future. My family sees this book as an opportunity to capture my father's stories and honor him. We want to preserve our history so that it can be shared with future generations, whether it is the Silverstein family or the whole Arch family, including its employees and customers. This book is dedicated to my father for his unrelenting determination, his focus on family values, and his strong work ethic. His vision and daily presence have made Arch Aluminum & Glass the phenomenal success it is today.

Acknowledgments

Many dedicated people assisted in the research, preparation, and publication of *Winning the Arch Way*. The development of a historical timeline and a large portion of the principal archival research was accomplished by research assistant Linda A. Firestone. Her thorough and careful work made it possible to publish a great deal of interesting information on the origins and evolution of this unique company. Senior Editors Anne Forsyth and Ann Gossy managed the editorial content, while Graphic Designer Ryan Milewicz brought the story to life.

The research, however, not to mention the creation of the book itself, would have been impossible without the dedicated assistance of Arch Aluminum & Glass executives and employees. Principal among these are Leon Silverstein, Jackie Audette, Roger Collette, Max Perilstein, Natalie Reed, Cathy Coleman, and Paul Parcell, whose affable guidance, careful editing, and willing coordination of records and individuals greatly assisted our research team. Special thanks are also extended to Samantha Beerman for her assistance in fine-tuning the manuscript for the Silverstein family.

All of the people interviewed—employees, retirees, family, and friends—were generous with their time and insights. Those who shared their memories and thoughts include: Donny Anderson, Gerald Ashley, Donna Aubuchon, Kevin Aubuchon, Paul Baskwell, Don Beaver, Gerry Boylan, Billy Britt, Ron Brock, Paula Coben, Douglas Cove, Richard Davis, Joseph Devine, Gabe Garcia, Les George, Garth Goodrich, Fred Hankerson, Mary Handi, Numa Handi, Don Hyde, Art Jackson, Richard Jaffe, Arnold Kramer, Mark Kuchenrither, Tony Lawson, Gene Lomando, Ray London, Bridgio Lopez, Orpha Lopez, Cliff Monroe, Alan Moses, Claire Murphy, Neil Murphy, Shawn Myers, Steven Perilstein, Don Powers, Paul Reifman, Ruth Reifman, Mike Rogers, Beani Silverstein, Bobby Silverstein, Leon Silverstein, Ricky Silverstein, Cindy Stocklin, Arnie Studenski, Dan Wagner, Dave Williams, and Jeff Ziesche.

Finally, special thanks are extended to the staff at Write Stuff Enterprises, Inc.: Jason Upright, executive editor; Elizabeth Fernandez, and Heather Lewin, senior editors; Sandy Cruz, vice president/creative director; Elijah Meyer, graphic designer; Roy Adelman, on-press supervisor; Abby Hollister and Christine McIntire, proofreaders; Mary Aaron, transcriptionist; Elliot Linzer, indexer; Amy Major, executive assistant to Jeffrey L. Rodengen; Marianne Roberts, executive vice president, publisher, and chief financial officer; Steven Stahl, director of marketing; and Tania Overby, bookkeeper.

In 1898, Harris Perilstein, along with his wife Fagie Mollie, opened a small glazier's shop at 519 South 6th Street in Philadelphia, Pennsylvania. After Harris' death, Fagie successfully managed the business with the help of their children. *(Photos from* Golden Jubilee, *a History of H. PERILSTEIN.)*

CHAPTER ONE

STEPPING TOWARD THE FUTURE

1930 – 1957

He was bright and there were things and places he wanted to be.

—Beani Silverstein, regarding her father-in-law Sydney Silverstein

ROBERT "BOBBY" SILVERSTEIN founded Arch Aluminum & Glass Company, Inc., in 1978. From its humble beginnings, with a single truck and a handful of employees, Arch has become one of North America's largest architectural glass and aluminum organizations. After 30 years of quality service, this American, family-owned business has 32 branches nationally, employs more than 2,500 people, and boasts assets of nearly $200 million. To understand why and how this company with a small-store, homespun culture prospers, it is necessary to go back to the early 20th century.

The story begins during the Great Depression. The stock market crash of 1929 left the country reeling. Foreclosures were the norm; unemployment was skyrocketing; long lines to get a bowl of soup characterized a day in the lives of many Americans. The jobless rate reached more than 4 million.

Yet, despite these hard times, new technologies and the production line symbolized the beginning of a new era. The country needed housing, and monumental buildings like the Chrysler Building and the Empire State Building were rising up. The headlines of a New York City newspaper reporting the construction of Rockefeller Center read: "A project of hope in the midst of despair is rising in the heart of Manhattan."[1]

The modernist movement grasped the imagination of architects around the world. And, although the United States was in the midst of a depression, modernism served as an expression of hope. Glass, as a structural part of a building, encompassed one of the major shifts in this new architectural perspective. Natural light and a sense of space characterized this new approach.[2] Advancements appropriate to the Industrial Revolution provided architects and designers with the ability to create structures in the spirit of the new age.

Centuries of hand-blown glass production were about to come to an end. Until the late 19th century, glass production remained slow, required great craftsmanship, and the finished product was generally considered a luxury item. Between 1904 and 1920, the window glass industry transformed itself with the assembly line, forever altering the manufacturing and handling of glass. By 1928, hand-blown window glass production had all but disappeared in the United States.[3]

Sydney Silverstein sitting at his desk at West Penn Mirror Company. His office door stands open in the background. This open-door policy began at H. Perilstein more than 100 years ago and continues at Arch Aluminum & Glass today. This family-run business takes pride in maintaining an organization that encourages dialogue at all levels with top management.

The use of the assembly line allowed manufacturers to produce window glass at a faster pace, and this increased efficiency provided the capability to produce specialized glass products. These specialized products, including sheet glass, tempered glass, and safety glass, became the new standard. The building industry quickly embraced these unique products for architectural purposes.[4]

Having the ability to produce a range of new, specialized glass inspired architects. Taking the modernist view on natural light and space gave rise to a whole new creative line of structural glass products, including glass building blocks, reinforced plate glass, and pigmented structural glass. These novel glass products revolutionized the building business.[5]

Metal and glass buildings initiated the modernist architectural movement. Buildings like the spectacular Crystal Palace, constructed in Hyde Park, London, for the Great Exhibition in 1851, with more than 1 million feet of glass, relied upon cast iron for the skeletal structure.

Not until 1886, when Charles Martin Hall invented a new process for obtaining large quantities of aluminum, did the production of aluminum come into play. From this point forward, aluminum became a desired metal for household items and for framing windows, doors, and buildings. By 1933, 13 U.S. companies banded together to form The Aluminum Association. World War II created a boost in aluminum production, and to meet the demand, the government financed and built several aluminum plants.[6]

A Growing Success

In this environment, H. Perilstein, a family-owned glass business, thrived. Started in 1898, H. Perilstein continued to prosper during the boom of the 1920s and increased its range of influence during the depression years. Opening new plants in several states, H. Perilstein flourished, while many businesses around it collapsed.

Harris Perilstein, founder of this accomplished company, grew his business based upon a simple concept: Take care of customers and employees, and a business will succeed. Although seemingly obvious, Harris' fundamental rule remained a core precept in the business practices for the members of two families, across several generations, engaged in the glass and aluminum trade.

The glass industry, typical of the times, attracted skilled craftsmen from Europe, many of whom were Jewish immigrants. Among them, the Perilsteins maintained strong Jewish religious convictions that included the traditional belief of helping others. Consequently, the Perilsteins donated their hard-earned money to various charities and Jewish orga-

Right: H. Perilstein switched to a faster means of transportation in 1923 by adding a fleet of trucks to its equipment line.

Below: In the first decade of the 20th century, H. Perilstein used horse-drawn wagons to deliver glass. Harris' oldest son, Nathan, fed, groomed, and handled the horses. Horse-drawn wagons allowed H. Perilstein to deliver glass to its customers in a timely fashion. *(Photos from* Golden Jubilee, *a History of H. PERILSTEIN.)*

Above left: At H. Perilstein, a 35-year skilled veteran cuts glass.

Above right: A craftsman at H. Perilstein bevels glass by grinding it on an abrasive wheel.

Below left: Early glass manufacturing required special handiwork. Here, a workman makes a mirror by pouring a chemical liquid on a table that maintains a constant temperature for even drying. Today, precision machines handle the burden reserved in the past for skilled glaziers such as these artisans. The modern techniques allow for uniformity in products and mass production.
(Photos from Golden Jubilee, *a History of H. PERILSTEIN.)*

nizations. Bobby, whose father worked for the company, recalled:

> *It was a family business, and it was a big family. They gave almost all their money to charity. They really never had a lot of money, the family, that is. A lot of people thought they were a very wealthy family, but they weren't.*[7]

Beani Silverstein, Bobby's wife, added: "They had no furniture. They gave it all away to the synagogue, to the family, to whomever. And, anybody that worked for the Perilsteins, regardless of your position or responsibility, you got the same amount. The Perilsteins took care of everybody."[8]

Harris first arrived in America in 1898 with only a diamond cutter in his pocket. After he fell ill and died, his wife Fagie Mollie, who had worked closely with him in the business, successfully carried on the family enterprise with the help of her children.

Striving to stay current with new technology and market trends, the H. Perilstein Company built and dedicated a new building in the memory of Harris Perilstein. That building, erected in 1929, incorporated both steel and concrete, and was fireproof, light, and airy.[9] This building, and others that would follow, reflected an approach to conducting business that both Sydney Silverstein (Bobby's father) and Bobby shared: Always innovate, and don't be afraid to take calculated risks.

Over the years, Harris and Fagie Mollie had nine children. Active, bright, and athletic, the youngest of the Perilstein children, Dora, graduated from Temple University. During her college years, she played field hockey and studied to be a physical education teacher.

Dora's outgoing personality attracted Sydney Silverstein, a smart, good-looking, ambitious young man who worked as a bookkeeper, earned a good salary, and went to night school. He grew disillusioned with night school, but not with Dora, and it was not long before they married.

A native Philadelphian, Sydney was one of three children. Although he spoke little of his youth, it is known that his father owned a poolroom, and that Sydney was his mother's favorite. Perhaps it was the ambiance of the poolroom that influenced the cultivation of Sydney's suave manner, motivation, and good looks, which he used to advance himself. One thing is certain: Sydney's love of sports, something that would become significant in his future, would provide a strong connection between himself and his family-to-be.

In 1947, H. Perilstein formed and incorporated a new company called Extrudite Metal Works, which produced storefront aluminum at a time when few such plants existed. With this acquisition, H. Perilstein now fabricated both glass and aluminum. Taking advantage of this opportunity, Sydney learned about the burgeoning aluminum industry. Louis Perilstein, Sydney's brother-in-law, would design the first line of aluminum products at this plant, helping H. Perilstein grow into a one-source shop for both glass and aluminum products.[10]

Once Sydney and Dora were joined in marriage, Dora and other members of the Perilstein family wanted Sydney to join the family business. He initially resisted the proposition, and he continued to work as a bookkeeper for two additional years. Eventually, however, in 1932, he succumbed to the pressure of the family, took a 50 percent reduction in pay, and went to work for the Perilsteins. As a family member, Sydney did not receive any special treatment on the job. He started at the bottom of the ladder, learning every aspect of the burgeoning aluminum and glass business.

Joining the staff of H. Perilstein provided Sydney with a new career direction, and one in which he excelled. His rise to stardom within the industry did not take long, and the Silversteins would further establish and enhance the one-source shop concept through four generations in the architectural product business.

An Early Warning Sign

Dora and Sydney started a family of their own with the arrival of Paula in 1931, followed four years later by Robert Harris (known to all as Bobby), and finally Natalie in 1938. Sydney worked hard and traveled frequently, striving to satisfy his insatiable appetite for success. Work encompassed most of his time and, as a result, Sydney paid little attention to family issues or his personal health and well-being.

Overworked, overweight, and a heavy smoker, Sydney experienced his first heart attack at the age

This picture, taken in the late 1940s, illustrates the manpower required to move a large sheet of glass. Today, machines help minimize accidents and possible damage to sheets of glass. *(Photos from* Golden Jubilee, *a History of H. PERILSTEIN.)*

of 38. When traveling from Philadelphia to Pittsburgh, and planning to head to Ohio next, Sydney visited Max Perilstein, his cousin, and Max's son-in-law, Paul Reifman. Paul's wife, Ruth, was a nurse and recognized Sydney's ill health. She urged him to go to the hospital for a checkup, but Sydney resisted the suggestion, claiming urgent business in Ohio. Fortunately, Paul persisted and drove Sydney to a hospital where his wife worked. Sydney was admitted and diagnosed with having recently suffered from a severe heart attack.[11] He stayed in the hospital for six weeks, and Dora brought their children to Ohio to spend quality time with Sydney. Many years later, Sydney's disregard for his own health would lead to a fatal heart attack.

Despite Sydney's single-minded focus on work and his strong desire to become highly successful, his children adored and respected him. To accommodate the family, Sydney reserved weekends for family outings. Bobby recalled:

> *We were a close family. We used to go to sporting events. We had season tickets to the professional football games at that time. My dad traveled a lot for the company, but that's probably why we went to sporting events on the weekends.*[12]

Sports would be the glue that bonded Sydney and Bobby together in the years to come. That love of sports would be passed down to Bobby's children and grandchildren alike.

Outside the family, Sydney's polished manner, easy communication style, motivation, and positive relationships with customers would serve him well. Always well dressed, well groomed, and well spoken, and, typical of the times, poised with a cigarette in his hand, Sydney took great pride in his appearance and his public persona.

Using his meticulous appearance and manners, he formed durable business and social relationships with many clients. "Anything Sydney indulged in, anything he did, it was like the saying 'hands of gold.' That's a good description of Uncle Syd," Paul said. "He was loyal to his friends, his vendors, anybody that he really was in contact with."[13] Ruth added, "I don't think you could find anybody that could really say a bad word about Uncle Syd."[14]

Sydney intuitively understood the value of building relationships in business, and passed that important concept on to his son and future business associate, Bobby. In the years to come, Bobby worked closely with his dad, but never became his partner. He remained a salaried employee the entire time he worked for his father.

A handsome couple, Sydney and Dora relax together on a vacation early in their marriage. Sydney originally resisted Dora's urging to join the family glass business, but, eventually, he agreed to work for his in-laws, and his life's career began.

The Making of a Dream

Notwithstanding his success at H. Perilstein, Sydney yearned to move to Florida. In 1936, he and his brother-in-law, Louis, took a trip down south to investigate business possibilities. With keen foresight, Sydney anticipated South Florida's booming business environment, with expansive building prospects and a great need for glass and metal products. Despite the beckoning business opportunity,

The Perilstein family owned a large home at the New Jersey shore. In-laws and other family members regularly enjoyed vacationing there, spending lazy summer days at the beach and fun-filled afternoons strolling along the famous Atlantic City boardwalk. Above left: Sydney, son Bobby, daughters Paula and Natalie, and wife Dora. Above right: Sydney, Paula, Natalie, Bobby, and Dora on the boardwalk.

Dora felt torn about leaving her brothers, sisters, and mother, and she adamantly refused to move.

For the time being, Sydney and Dora remained in Pennsylvania, where in 1945, Sydney organized the setup and opening of a new state-of-the-art mirror and tempering glass facility called West Penn Mirror, Inc., in West Pittsburgh. It took a year to ready the plant. The grand opening occurred on December 15, 1946, and took on all the aspects of a joyous family outing.[15]

Bobby, an active 10-year-old at the time, vividly remembered the occasion: "The day before the celebration picnic, I was playing football and got thrown onto a broken glass bottle lying in the street. I nearly cut my hand off. They took me to the hospital."[16] It seemed as if Bobby would miss the big event. Luckily for Bobby, H. Perilstein had an infirmary at the plant and a contract with a doctor to be on call to tend to any employee who had been injured in the course of a day's work. This doctor planned on attending the celebration, so Bobby's parents allowed him to join the two railroad cars full of family, suppliers, and customers traveling to the ceremony from Philadelphia to Pittsburgh.[17]

With the injury behind them, the family enjoyed the company's inviting atmosphere, which would continue throughout Sydney's career, and again when Bobby opened Arch Aluminum & Glass. The importance of a homespun family environment, one that treats workers as equals, led to a unique company culture at Arch and fostered intense employee, customer, and supplier loyalty through the decades.

The grand opening at West Penn Mirror helped communicate the value of the modern production line, which utilized 10 different operations to manufacture its products. Sydney grasped the economic benefits of the assembly line, including its ability to reduce the time and cost of producing products. During the next five years, Sydney continued to position himself and his company strategically as a forward-thinking organization within the glass and mirror industry.

Sudden, Unexpected Changes

For Sydney, life took a turn for the worse in the late 1940s. Dora died prematurely in 1949, leaving Sydney with an adolescent son and daughter, and one young girl. Sydney sold the house and moved the family into an apartment to escape the sad memories of losing his wife and the difficulties of maintaining the upkeep of the home and raising children as a single father who traveled frequently. Because Sydney placed a great importance on providing a good home for his family, he quickly found a new wife to provide for the children. Bobby recalled:

> *My father married a woman, a Jewish woman, [whom] he thought would be a good mother. We were living in an apartment then. It was maybe two years after my mother had died. My father ended up sending my younger sister to live with a friend of his so that she could attend the same school as she had prior to my mother's death. It really wasn't going along well at all, even after a month into the marriage. Then, soon thereafter, he moved out, and I stayed there with my stepmother.*
>
> *It was probably a week that I was there with just her, maybe two weeks ... [when] one day, I came home from school and opened the door, and there was nothing in the apartment. It was completely cleaned out, and my new stepmother was gone. My dad ended up getting furniture, and we [stayed] there.*[18]

Bobby survived the trauma of losing his mother and a stable home environment during adolescence. In later years, the loss of his mother would help Bobby comfort his 19-year-old wife, Beani, whose father was diagnosed with cancer and died quite suddenly. Bobby also learned, during these years, the uselessness of looking back and dwelling on the past. For him, the past was finished and the future was before him, and he maintained that attitude throughout his life.

While still in school, however, Bobby was an angry young man, and he released his aggression on the basketball courts. In high school, Bobby played on his junior varsity basketball team during his junior year and led the team in scoring. Sadly, Sydney never attended one of Bobby's games. "Unlike his father, when Bobby traveled, he would fly home just to coach a game or catch one of the children playing on a team," Beani recalled.[19] Bobby tried out for varsity his senior year, but he did not make the team because it was loaded with talent, including his friend Marvin Cravetz and a freshman sensation named Wilt Chamberlain. Rather than playing junior varsity as a senior, Bobby quit the team, but he never resented Cravetz, and down the road, they became business partners.

During Bobby's high school years, Sydney continued to work for H. Perilstein, rising to the position of vice president. Periodically, Bobby would come to work with his dad and catch a glimpse of the glass business.

In the summer of 1951, Bobby and Beani met in Atlantic City. Beani recalled:

> *The brother of my sister's husband knew Bobby from school, so that's how we met. I thought Bobby was very cute, very quiet, very, very handsome, and independent. He was very much on his own.*[20]

Around the same time, Sydney again remarried, this time to Margie Palie, a non-Jewish nightclub singer. As a religious family, the Perilsteins were displeased with this union, which led to friction between Sydney and the Perilstein family. In 1952,

Bobby graduated from high school in 1950. He disliked school and looked forward to pursuing a career after graduation.

Early in his career, Sydney Silverstein demonstrated a proven ability to understand market trends and embrace new technology. In 1946, Sydney organized, opened, and managed West Penn Mirror, Inc. The most modern plant of its type at the time, West Penn occupied more than 50,000 square feet of space. Sydney would maintain ownership of the company until 1955. *(Photo courtsey of Seavy's Studio.)*

WEST PENN MIRROR, INC.

A Primer on Glass

NATURALLY OCCURRING GLASS HAS existed since the beginning of time, and the earliest man-made glass objects found in Egypt and Eastern Mesopotamia date back to around 3500 BC. Until the mid-1800s, glassmaking remained a specialized art within the domain of skilled craftsmen. Not until the latter stages of the Industrial Revolution did the craft turn into industry. The necessity for the artist's skilled hands was replaced by the use of machines. Machine technology and the assembly line enabled the production and utilization of glass for commercial purposes.

Three principal raw materials make up commercial glass: lime, silica, and sodium carbonate. By mixing secondary materials with these raw materials, different types of glass are formed. Almost all commercial glass falls into three basic categories defined by various chemical processes used to create each type. These categories include soda-lime, lead, and borosilicate glass. Although only three categories of glass exist, numerous types of glass created by different chemical processes make up these categories.

- **Float Glass:** This process, introduced in 1959 by the Pilkington brothers of England, revolutionized the glass-production process. By pouring molten glass onto one end of a molten tin tub, the glass floats on the tin and spreads out along the bath, producing a smooth surface on both sides. Because this type of glass breaks easily into large jagged shards, it is not recommended for architectural use.
- **Sheet Glass:** Before the invention of float glass, sheet glass was created by dipping a leader into a vat of molten glass and then pulling it up so the film of glass hardened. This technique produced clear glass. However, variations in thickness resulted from small temperature changes during the hardening process. Even today, sheet glass can be found in the windows of some older homes.
- **Safety Glass:** Upon impact, tempered glass, a type of safety glass, crumbles into thousands of tiny pieces and falls to the ground, rather than shattering into sharp shards of glass. Tempered glass is often found in unframed structures such as showers and car door windows. Another type of safety glass called laminated glass remains intact when shattered. An inner layer of plastic bonds two or more layers of glass and produces a cracking pattern upon impact. Storefront windows and automobile windshields typically utilize laminated glass.

with Sydney's blessing, the Perilsteins bought Sydney out of H. Perilstein and provided him with the ownership of West Penn Mirror.

A Time for Change

Sydney would hold ownership of West Penn Mirror for a few more years. During this time, Bobby worked with his father, driving from Philadelphia to West Pittsburgh on Sunday evenings and heading back to Philadelphia on Friday nights.

While working there, Bobby observed that the seaming operators were only completing two-and-a-half pallets a day because they were performing at a slow pace. One morning, Bobby told them they could go home after they finished seaming three pallets to encourage them to work faster.

The seaming operators finished the pallets and left by lunchtime. But the other workers got upset and filed a grievance because they did not get to leave early.

After this incident, Sydney told Bobby, "Don't talk to the people on the floor. If you have an observation, let me know, and we can discuss it."[21] Years later, Bobby gave Leon, his oldest son, the same rules when he came to work for Bobby.

In 1956, Sydney sold West Penn Mirror to the Hamilton Glass Company. Ironically, this state-of-the-art plant never reopened. The Hamilton family experienced a series of tragic events and, as a result, the plant remained closed. Sixty years later, with broken windows, years of disrepair, and overgrown landscaping, this building remains a sad shadow of its promising beginnings.

After selling the business, Sydney worked at a variety of consulting positions to occupy his time but did not attempt to secure a permanent position, as he still dreamed of owning a business in the South. In 1957, the year Bobby married Beani, Sydney realized his dream. Perhaps not coincidentally, Sydney went down to Miami to buy a business at the same time Bobby and his new wife Beani honeymooned there.

While enjoying their honeymoon, Bobby and Beani received a call from Sydney. "If I had known he was going to be down here," Beani recalled, "I wouldn't have been too happy. And so he said, 'Let's go to the races.' He knew I'd like that."[22]

During the excitement of the races, Sydney informed Bobby and Beani of his own good fortune; he had just bought a glass business, Pritchard Paint and Glass, in Miami. Beani explained:

> *There was an unspoken understanding that Bobby would work for Sydney. Sydney didn't know if the business would be a success. But, if it didn't work, he said, "we'd squeeze grapefruits." It was unlikely we'd need to resort to that. Remember, Florida has hurricanes. When you have hurricanes, you need glass.*[23]

Making the decision to live and work in Miami proved difficult for the newlyweds. Beani's father was gravely ill at the time. Bobby and Beani deliberately moved up their wedding date by several months so that Beani's dad could attend. Besides, Beani felt torn about leaving her mother and sister. Despite her reluctance to move, Beani agreed because she loved and admired Bobby. Beani recalled:

> *Bobby was not like most young people. He had lost his mom early in life. He was very brave. During high school, Bobby worked for his uncle. He went back and forth to New York to pick up cabana shirts. After he graduated from high school, Bobby drove from Philadelphia to West Pittsburgh to work with his father. He did things at 18 most young people don't need or want to do. All he wanted was a jingle in his pocket, and he worked hard for it.*[24]

Under Sydney's guidance, and through Bobby's hard work, the new glass company in Florida would thrive.

Sydney used this truck to deliver glass within the Miami area shortly after starting his business. Within two years, his delivery area had extended across the state.

CHAPTER TWO

A Growing Enterprise

1957–1970

Sydney wanted more, and he got more. You respected him. That's what you did.

—Beani Silverstein

WITH THE PURCHASE OF A glass business, Sydney Silverstein would begin to fulfill his dream of becoming a successful entrepreneur. Slipping easily into his new role, he comfortably networked with the big guys in the industry and presented his fledgling business in such a way that people listened.

The unprecedented growth of Sydney's new company, exceeded only years later by the major expansion of Arch Aluminum & Glass, required careful attention to market conditions and the integration of ever-developing technologies. Changes in glass production in the 1950s, which allowed for the production of large sheets of flat glass, coupled with the increased ability to manufacture aluminum, ushered in the common building feature known as metal and glass curtain wall systems. These systems proved cost efficient, and Sydney capitalized on their distribution in his business.

To make the best possible use of the new technologies, and to distinguish his business entity from the old Pritchard Paint and Glass, Sydney renamed his company Architectural Glass Products and changed the focus from installation to distribution. He then tapped into the business relationships with customers and suppliers he had developed while working at H. Perilstein and West Penn Mirror. Paul Parcell, who met Bobby and Sydney in 1961 while working as a self-employed glazier, reflected upon the unique relationships the Silversteins were able to cultivate in the business. "I had a good feeling about Sydney. When my partner and I broke up the business, I had already come to know Syd. He helped me out and gave me open credit," Parcell recalled.[1] More than 30 years later, Parcell happily joined Arch Aluminum & Glass.

Employee Relationships

Today, Parcell loves coming to work because he enjoys working for the Silversteins, and he's not the only employee who shares this feeling. All 26 Pritchard Paint and Glass employees stayed on to work with Sydney and Bobby after the acquisition. Past employees, customers, and suppliers have always expressed a loyalty to the Silversteins. One of those early employees, Arnie Studenski, who worked with Bobby at Arch Aluminum & Glass until he retired, recalled, "I've always gotten along well with Bobby. He's always been good to me."[2]

These enduring relationships helped carry Architectural Glass Products and Bobby's future

At Architectural Glass Products, Sydney took off his jacket to help out in the warehouse. Above, he reviews inventory lists.

business ventures through many transitions. Personal stories of individuals involved in the business with Bobby help to illustrate the power of those relationships. For example, Roger Collette, a current employee who has worked for Bobby on and off since the late 1950s, described how he came to work for the Silversteins:

> *My mother worked for the William L. Bonnell Company, which was a supplier for Bobby. My mother had done a lot of favors over the years for him, routing the trucks and [providing] special pricing to favor the account. He always appreciated it. He'd take her to nice places for lunch. One time, she asked him if he'd do her a favor and hire her son. That's how I started working for Bobby. My mother is still alive. She's 90 years old and still talks to Bobby on the phone.*[3]

To Bobby's credit, many current employees have either worked with him in some capacity over the past 40 years or forged long-term careers with Arch Aluminum & Glass as it grew. The unpretentious leadership at Architectural Glass Products, and now at Arch, fostered this impressive employee retention. Bobby views himself as, and relates to, the common guy, interacting easily with the men and women who work for him in both the administrative offices and in the warehouse. "We like common folks. We have a lot of people off the street doing jobs. They have learned the trade working their way up the ladder," Bobby said.[4] Bobby provides his staff with the opportunity to experience career growth at all levels, just as he did almost 40 years ago.

A Father's Example

Within a year of business, Architectural Glass Products distributed not only glass but also mirrors and aluminum doors. Architectural Glass Products sold these products throughout most of the state of Florida, and the company experienced rapid growth as a result of its expanded product line.[5]

THE ROAD LESS TRAVELED

ROGER COLLETTE WORKED FOR SYDNEY and Bobby at Architectural Glass Products and Aldora Aluminum for years. He started as a mechanic and moved into driving and making deliveries. Traversing the state back in the early 1960s proved quite different from the direct route available today. Alligator Alley, the only highway to cross South Florida, did not exist back then, so the trip across the state was slow and circuitous. Collette, who made the delivery run regularly, reflected:

> *I would take the long trips to the west coast of Florida. Those trips went from Miami across to Naples. ... This was before Alligator Alley. You drove across many back roads to Naples, and then you went north and worked your way up to Clearwater and Tampa. Finally, you'd drive across again. Sometimes, I'd work 22 straight hours dropping glass off to different glass shops, and metal to their shops. Some of those shops are still over there, generations later.*[1]

And, to Bobby's credit, many of these same shops utilize Arch Aluminum & Glass today.

When Sydney bought Pritchard Paint and Glass, all 26 of the workers stayed on. Second from the left stands Roger Collette. He has worked for Bobby off and on since 1958. Today, Roger works as Bobby's assistant.

Above left: Workers at Architectural Glass Products unload products from a truck at the warehouse, while Sydney (center) surveys the area.

Above right: In the early years of Architectural Glass Products, Sydney only distributed glass. In the warehouse, each sheet of flat glass was stored in an individual slot for safekeeping.

As the company grew, Bobby learned through Sydney's example. "My father told me, 'You'll sit at a desk in my office and you'll listen. [If] you don't like what I have to say to other people—you'll still sit and listen. We'll talk about it afterward,'" Bobby reminisced.[6] After a year of doing just that, Bobby took charge of warehouse operations. Bobby's insight and his ability to relate to the men and women in the warehouse served him well over the years. "You know, if you have the favor of the workers, it's much easier to succeed than if you have to fight them," Bobby said.[7] Neither Architectural Glass Products nor Arch ever unionized.

To this day, Bobby maintains an open-door policy, where employees are encouraged to come to him confidentially with problems or concerns. He explained:

> *I always felt that if people were going to work for me, I wanted them to feel comfortable working for me. I didn't want them to see me as their boss. I was always the boss. They knew that. But I didn't want them to think of me that way because if someone came in and had a problem, I wanted to hear it. I didn't want them to cover it up. Even today, employees will come in and talk to me. If they tell me things that shouldn't be told, it's not going to go anyplace else but me. I want them to always feel that they could talk to me. They can't talk to me if I'm a "mister." They just can't.*[8]

In addition to strong employer–employee relationships, Architectural Glass Products succeeded as a result of a simple operational philosophy. Bobby, ever the pragmatist, explained: "We were strictly bread and butter, we made simple things. If you stick with that, you could grow with that, because you could give better service."[9] This philosophy would later help Bobby launch Arch Aluminum & Glass and foster its growth into a national corporation.

And Baby Makes Three

During that first year in business, as Architectural Glass Products grew, so did Bobby and Beani's family. Under some pressure from the traditional Sydney, they decided to have a baby. In 1958, they welcomed their first child, Donna, into the world.

Despite the business success and Bobby's newly acquired responsibilities at the company,

Sydney paid Bobby a meager salary, and Bobby's new family struggled financially. This time period proved difficult for Beani. She felt lonely and cramped in their one-bedroom apartment. She missed her family, and she missed Philadelphia.

"We didn't know anybody. I wasn't even allowed to have any pets," Beani mused. "Margie [Sydney's wife] found our apartment. We had very little money. Bobby was making $90 a week. We had this apartment, a maid, and a cabana that Margie had gotten for us, but we didn't have money to park the car. The apartment, the maid, the cabana, that was all Margie thought was important."[10]

With the addition of Donna, the apartment seemed even more stifling. Bobby and Beani realized they needed a bigger home and contracted to build one. Sydney believed Bobby could not afford a new home, and he strongly discouraged him from building one. Beani recalled: "It was the only time I ever saw Bobby blow up at his father. Bobby told Sydney, 'When I can afford [the house], I won't need it. So I'm doing it now.' And the builder lent us the money. Sydney wouldn't lend us the money. No, no, you have to creep before you crawl—that was Sydney's belief."[11]

The young family benefited from the move to a larger home in a young, growing neighborhood. However, Bobby still occasionally grew impatient with his financial situation. "There were times he thought he could do more. Make more money. We struggled. I used to talk to my next-door neighbor all the time. Her name was Dolores. We would talk and we would say if only our husbands made $10,000 a year. I remember this like yesterday. But, then I was happy, because we lived close to all those young people, and then things got better. I needed my roots," Beani remembered.[12]

Their first son, Leon, was born in 1960, shortly after their move. Beani described those early years as a time filled with both joy and struggle.

Bobby's love of cars got him in a little trouble when he surprised Beani with a new MG. Beani did not appreciate Bobby's extravagant purchase, because, at the time, the family could barely afford their living expenses. But they would work through it, and Bobby would keep the car.

Acquiring Product Lines

While Beani and Bobby settled into their new home, Sydney continued to expand his business and pursue old contacts from his years in the glass business up north. He contacted Abe Rabinowitz, a former employee of H. Perilstein who had relocated to Miami Beach and owned Adams Glass Company. "We went to see him. That's the reason we probably even went into business down here. Also, Abe's son-in-law, Harold Dudoff, sold aluminum for the second-largest metal manufacturer in the country, called Amarlite. He worked as an export agent and helped establish our connections in Cuba, the Caribbean, and South America," Bobby recalled.[13]

Abe and Harold introduced Sydney to the owner of Amarlite. "My father related very well to this individual and they became friendly. That's how we ended up selling their products," Bobby explained.[14] The manager and Sydney worked out an exclusive deal for Architectural Glass Products to sell Amarlite's aluminum door frames throughout Florida—a tremendous opportunity for Architectural Glass Products—even though Amarlite retained a few of its biggest Florida clients as house accounts. With this new association, Architectural Glass Products became the leading

South Florida distributor of aluminum commercial doors, and the business continued to prosper.

Planning to expand the product line even further, Sydney decided to add mirrors to his inventory. To accomplish this endeavor, he approached an old friend of his at Carolina Mirrors, whom he knew from his years at H. Perilstein, and they struck a deal that provided Architectural Glass Products with exclusive rights to distribute Carolina Mirrors in Florida.

At this point, Architectural Glass Products distributed glass, aluminum doors, and mirrors, providing customers with a one-stop shop capability. With its three-tier distribution model in place, an approach that continues at Arch today, Architectural Glass Products established a solid customer and supplier base. Three of the largest glazier contracting companies in the state—Adams Glass, Tropical Glass, and Collins Glass, all located in Miami Beach—routinely purchased products from Architectural Glass Products. Sydney and Bobby valued these loyal customers. Bobby explained:

> *Sometimes they paid a little slow, but my philosophy ... is trust. If I trust you, then I will carry you.*

Right: Sydney, in his usual dapper dress, attended glass, mirror, and aluminum trade shows to network with other businesspeople and keep abreast of new products and technology in the industry.

Inset: In 1963, Sydney attended a large trade show in Los Angeles, California. Here, he stands with longtime acquaintance Ralph Buchan. During this time, Sydney held exclusive distribution rights in Florida for Carolina Mirrors and Amarlite doors and window frames.

Below: From left, Sydney, Margie, Beani, and Bobby frequently went out to dinner with clients. On these occasions and at the trade shows, Bobby wore a jacket and tie.

I get better prices. I'd rather gamble on you than sell cheap and be paid right away. I would rather make the bigger margin and gamble than take the big account and hope nothing goes wrong.[15]

Bobby and Sydney secured and retained customers and suppliers by building relationships with the clients. Bobby and Beani socialized frequently, but, as Beani explained, "Everything was business. That was our social life. Everything was going out with customers. We socialized, but it was all business."[16]

Sometimes, those business outings involved attending sporting events. Sydney and Bobby bought blocks of tickets and invited clients along. They even rented buses and transported their business associates to the Orange Bowl Classic for years.[17] This tradition of entertaining clients lives on today at Arch Aluminum & Glass, where the company hosts various events for its customers and employees.

Waging a War

Despite its loyal following and strong customer base, Architectural Glass Products' success became threatened when the management of Amarlite changed in 1963. Sydney's old contacts began to retire, and shifts in the organization occurred. The new management decided to relinquish Architectural Glass Products' right to maintain distribution control of its products in Florida.

Rather than lose his footing in the aluminum door business, Sydney challenged Amarlite to a battle by revving up the competition and opening his own aluminum door manufacturing and commercial extrusion facility. He rented the vacant lot across the

Above and below: Sydney bought his own trucks to transport the products distributed from Aldora and Architectural Glass Products.

Below left: In response to Amarlite's move to cut Architectural Glass Products out of the aluminum business, Sydney and Bobby built an aluminum plant across the street from Architectural Glass Products.

One of the many modern entranceways created by Aldora as depicted on the cover of an early catalog. Architectural Glass Products provided the glass and Aldora provided the metal, enabling the architect and general contractor the ability to one-stop shop.

street from Architectural Glass Products and wasted no time building his own aluminum plant. Sydney decided to call the new operation Aldora Aluminum after his first wife, Dora, the mother of his children. Arnie Studenski helped design and set up the workspace for the new business that would manufacture simple aluminum doors.

To retaliate against Sydney's challenge, Amarlite threatened to sue Architectural Glass Products for patent infringement on one of its door designs. Sydney disarmed his competitor by inviting the management of Amarlite to visit the Aldora plant. Sydney proudly showed them the unique design of his door. He had intentionally created a door similar to Amarlite's door, but with slight design variations to avoid any patent infringement. As a result, Amarlite had no case, and the management rescinded its threat. This ability to act quickly and to take a liability and turn it into an asset would be repeated three decades later through a major growth initiative at Arch Aluminum & Glass.

With the assault by Amarlite thwarted, Sydney and Bobby faced yet another challenge. Thefts occurred at both of their businesses. To prevent further loss of inventory, Bobby hired a security company to watch over the businesses, and the company brought in an evening guard with dogs to patrol the grounds. When the thefts continued, Bobby fired the security company. Soon after the dismissal, a fire mysteriously occurred in one of the buildings, forcing Sydney and Bobby to close up shop for almost six months to take care of extensive repairs. To this day, Bobby suspects the involvement of the security guard in the criminal activity.

No Price-Cutting

Undeterred by this setback, Architectural Glass Products continued to grow as a result of its steady customer and supplier base. Carolina Mirrors, American Safety Glass, and William L. Bonnell, in conjunction with Florida's biggest glazing contractors, Adams Glass, Tropical Glass, and Collins Glass, kept Architectural Glass Products and Aldora Aluminum busy and prospering. "We just grew our aluminum line. We added different products in the aluminum field. We were always a fairly aggressive company," Bobby explained.[18]

However, Sydney and Bobby did not cut prices to compete with or win over customers. Bobby explained:

A TESTAMENT TO QUALITY SERVICE

THE CUSTOMERS AND SUPPLIERS LISTED BELOW HAVE REMAINED LOYAL TO Bobby and Arch Aluminum & Glass since the early days of Architectural Glass Products and Aldora Aluminum.[1]

- AAA Glass Service (Washington, North Carolina)
- Acme Glass (Orlando, Florida)
- Acme Glass (Tampa, Florida)
- AlumiGlass (Boca Raton, Florida)
- American Glass (Hialeah, Florida)
- B & B (Lake Worth, Florida)
- Bailes (Melbourne, Florida)
- Beach Glass (Fort Myers, Florida)
- Bill's (Naples, Florida)
- Brandon (Brandon, Florida)
- Britt & Tilson (Asheville, North Carolina)
- Brook's Glass (Melbourne, Florida)
- Charlotte County (Port Charlotte, North Carolina)
- City Glass (St. Petersburg, Florida)
- Coral Gables Glass (Miami, Florida)
- Custom Glass (Pompano Beach, Florida)
- Davie Glass (Davie, Florida)
- Frank Burkart (Marietta, Georgia)
- Glass Master (Palm City, Florida)
- H.B.S. (Vero Beach, Florida)
- Keating Glass (Orlando, Florida)
- Lee & Cates (Jacksonville, Florida)
- Miami Engineering (Miami, Florida)
- Midway Glass (Clearwater, Florida)
- National Glass (Miami, Florida)
- Nilsen (Sarasota, Florida)
- Oneco Glass (Oneco, Florida)
- Pittsburgh Paint (Winter Park, Florida)
- Pompano Glass (Pompano Beach, Florida)
- Premier Imports (Nassau, Bahamas)
- Punta Gorda (Port Charlotte, North Carolina)
- Ridge (Lake Wales, Florida)
- Seminole (Pompano Beach, Florida)
- Smokey's (Bradenton, Florida)
- Tom Winner (Rockledge, Florida)
- Tropical Glass (Miami, Florida)
- Twin City Glass (Miami Beach, Florida)
- West Tampa (Tampa, Florida)

Neither my dad nor myself ever ran a business where we were going to be the low-cost seller. We might make it for less, but never undercut the price. Price wasn't our focus. We never went out on the street to try to cut a customer's price just to get the business. Even today, we don't do that. If you're just going to cut people's prices, in the long run, you're not going to make anything. We base our profit on service, and that's how we grew, and that's why we made money.[19]

A New Place to Work

With hard work and positive business relationships, Aldora and Architectural Glass Products continued to expand. By the mid-1960s, the businesses needed more space. The original building space in Miami for Architectural Glass Products measured 15,000 square feet. In addition to housing aluminum doors, glass, and mirrors for distribution, the companies now also distributed aluminum and fabricated glass. And as product lines grew and production increased, so did the number of employees and the sales force. Within three years, Architectural Glass Products and Aldora Aluminum increased the useable building space to a total of 30,000 square feet.

Sydney, never timid about taking risks, decided to build a plant twice that size. In 1965, Sydney moved Architectural Glass Products and Aldora Aluminum to an area along the Palmetto Expressway called Medley. This move proved strategic in two ways: The family now owned its own massive

facility, and both companies were housed under one roof.

After the new facility was fully operational, Sydney turned his sights to the next stage of development—out-of-state expansion. With the help of his friend, Marty Udenfriend, Sydney opened a small aluminum business as a test market site in Gloucester City, New Jersey. Udenfriend owned a business specializing in custom-insulated units for refriger-

Above left and right: These two photos illustrate the large area that would house the newly constructed Architectural Glass Products and Aldora Aluminum plant. This area was double the size of the space originally rented in 1957, utilizing a full 60,000 square feet.

Below: Sydney looks happy in his spacious new office, behind his new desk. The new facility in Medley had beautiful offices and a large waiting room.

Above left and right: This series of photos depicts the equipment used for tooling aluminum windows and door frames. Men placed the material on racks then cut it to size. After cutting the metal, it was packaged and racked near the truck entrance before being loaded onto trucks.

Left: In the late 1960s, workers packaged glass sheets, crated them in wood boxes, and stored them in the warehouse until shipment.

Below left and right: Dating back to the opening of West Penn Mirror in Pennsylvania, Sydney and Bobby celebrated holidays with their employees, customers, and suppliers. The Silversteins invited entire families to their gatherings to maintain the homespun quality so important to their business.

MEMORY LANE

As a young boy, Leon enjoyed visiting the family business and offering suggestions to improve its operations. From these very first experiences, Leon knew he would someday lead the company. He remembers frequently going to work with his dad. Even then, Leon focused on improving productivity and efficiency:

I would go there, and I would build forts in the warehouse. I still remember where they would leave lots of corrugation. They had tons, just mounds of it, and a lot of it was for the Howard Johnson accounts and all that. Arnie Studenski would bring his son to work as well. We used to play there the whole day. We'd go to the break truck at lunch. Then, we'd get our food, and we'd bring it back in there, and we'd work on the fort. When we weren't in the fort, we would work in the stockroom. The stockroom was more fun and games. I would count screws and nuts, and I would weigh things. I loved weighing things on the scale. So that was what I did. I also went to the Crandon Park Zoo all the time with the truck drivers because we would deliver, and afterward we'd stop at the zoo.

I also remember sitting on my grandfather's lap in his office and making recommendations. My father and my grandfather had an interesting office setup. They had an office on each end and a big adjoining conference room. I would run back and forth, and I would just hang out often with my grandfather. He always gave me money. I came to work, so he paid me. I'd walk the plant with him, telling him everything they were doing wrong in the back, at my age. So that was my early work experience—even back then I enjoyed it.

Another thing I remember as a kid is going to watch my dad play basketball. My dad loved playing basketball. He played into his late 30s. I remember as a kid, five, six, seven, eight, nine, 10, all the way up until I was probably 13 years old, I remember he always had a Sunday game. So I would go with him to his Sunday game and watch him play. He was a maniac. Aggressive. Physical. He didn't want to lose.[1]

ation there, and he offered Sydney limited space within his building to open an aluminum operation.

Within two years, Sydney's business had outgrown the small space and, in 1970, Aldora Aluminum moved its Gloucester operation to Pennsauken, New Jersey, and added Architectural Glass Products to the production line.

Sydney sent his son-in-law, Jerry, who had worked for him in Miami, to manage the new plant. Because Jerry had less experience than Bobby, his promotion created some tension in the family. "Bobby is kind of a stoic person. When Bobby's brother-in-law was getting more perks than he was, he really didn't feel Jerry deserved it," Beani recalled. "But, Jerry was older. Bobby wouldn't verbalize it. He just got himself sick over it. He developed an ulcer then. He showed such respect for [his] father."[20] Always the diplomat, Bobby attributes his father's actions to the fact that Jerry was older and had a large family to support.[21]

Instead Bobby focused on further developing the customer base, while Sydney continued to strategize about expanding his businesses into other states. In 1970, a powerful national company would approach Sydney with an offer he couldn't refuse. His life-altering decision would drastically change the course of events for the father-and-son enterprise.

In 1973, Bobby Silverstein (third from right) joined Chromalloy to manage its building materials division.

CHAPTER THREE

THE CHROMALLOY YEARS

1970 – 1978

I trust everybody, I treat everybody like we're equals. But when I find out they're not treating me right, that's the end.

—Bobby Silverstein

DURING THE LATE 1960s, A publically held company based in St. Louis, Missouri, called Chromalloy, was rapidly acquiring smaller companies to diversify its holdings. Chromalloy targeted glass and aluminum businesses in an effort to become a national leader in the industry. It purchased H. Perilstein, the company founded by Sydney's father-in-law, and the first glass business at which Sydney worked. Next, Chromalloy acquired one of H. Perilstein's biggest competitors, Hires Turner, taking over four of its five plants in the east. Finally, it bought Reuben Glass, based in Texas and New York.[1]

In a matter of months, Chromalloy had secured a good portion of the East Coast glass and aluminum trade. Then, in 1970, the company's executives struck a deal with Sydney to buy Architectural Glass Products and Aldora Aluminum Products. The specifics of the deal are unknown, but it is acknowledged that Chromalloy provided Sydney with company stock options and an offer to head its new glass division. At the time of the arrangement, Sydney's companies grossed approximately $20 million a year.[2] Even so, he jumped at the opportunity to sell his own businesses and become a major player in a large corporation.

Despite the fact that Sydney worked closely with his son Bobby, Sydney did not include him in the business negotiations or the final decision-making. Rather, Sydney and everyone else assumed that Bobby would stay in Florida and continue to oversee Architectural Glass Products and Aldora Aluminum Products operations, now under Chromalloy's ownership.

As a manager at Chromalloy and leader of the new glass division, Sydney traveled regularly from Florida to King of Prussia, Pennsylvania, to oversee the different glass facilities, including those of H. Perilstein. Unfortunately, some of the veteran employees of H. Perilstein resented Sydney's new position and his power as their new boss. Old resentments rekindled and tensions grew. Nevertheless, Sydney immersed himself in his work. He worked diligently, and traveled extensively, while succeeding in his cherished new job. Just prior to that time, Sydney had stopped smoking and felt better, but as he got caught up in his business activities, he started lighting up again and once more paid little attention to his health.

About six months later, after a three-day meeting at Chromalloy's headquarters in St. Louis, Bobby recalled, "I came home Saturday. My father came home on Sunday. When he got home he

This Chromalloy Aldora trade show booth, from the early 1970s, showcases some of the products, including doors and glass, distributed by the company at the time.

called me. We discussed the football games and all. Later that night, my stepmother called me and told me my dad had been taken to the hospital."[3] Sydney had suffered his second heart attack at the age of 62. He passed away in the hospital a few days later.

New Hierarchy

After Sydney's death, Bobby remained in Miami working at Architectural Glass Products and Aldora Aluminum Products for about two more years. Then, one day, some of Chromalloy's hierarchy, including the chairman of the board, came to Miami to see Bobby. Bobby explained:

> *We got into a car and started heading over to a fancy club in Miami Beach. We were riding around talking. Then the chairman of the board asked me, "Where did you go to college?" and I told him, I didn't go to college. He said it again, "Where did you go to college?" and I answered the same way. He asked me a third time, and this time I knew how he wanted me to answer. So finally, I said that I went to the University of Pennsylvania. The Chromalloy executive said, "That's where I thought you went." We had dinner with some other people, the types you would see in the movies like* Goodfellas, *and that's when they asked me if I wanted to move and work up north for Chromalloy.*[4]

The new job would require Bobby to relocate and start at his new position within a matter of days. His responsibilities and power of influence would expand considerably. Bobby would not only manage the newly built Chromalloy plant in King of Prussia but also the H. Perilstein facilities and other businesses Chromalloy had bought. "The area was like the hub for the aluminum and glass industry," Bobby said.[5]

By this time, the Silverstein family had grown to include Cindy, born in 1963, and Ricky, who came into the world four years later. When considering the offer, Beani recalled, "Donna was going to go into high school, Leon into junior high, Cindy into elementary school, and Ricky was just starting first grade. So it was a good time to move, a really good time to move. And I think Bobby was happy

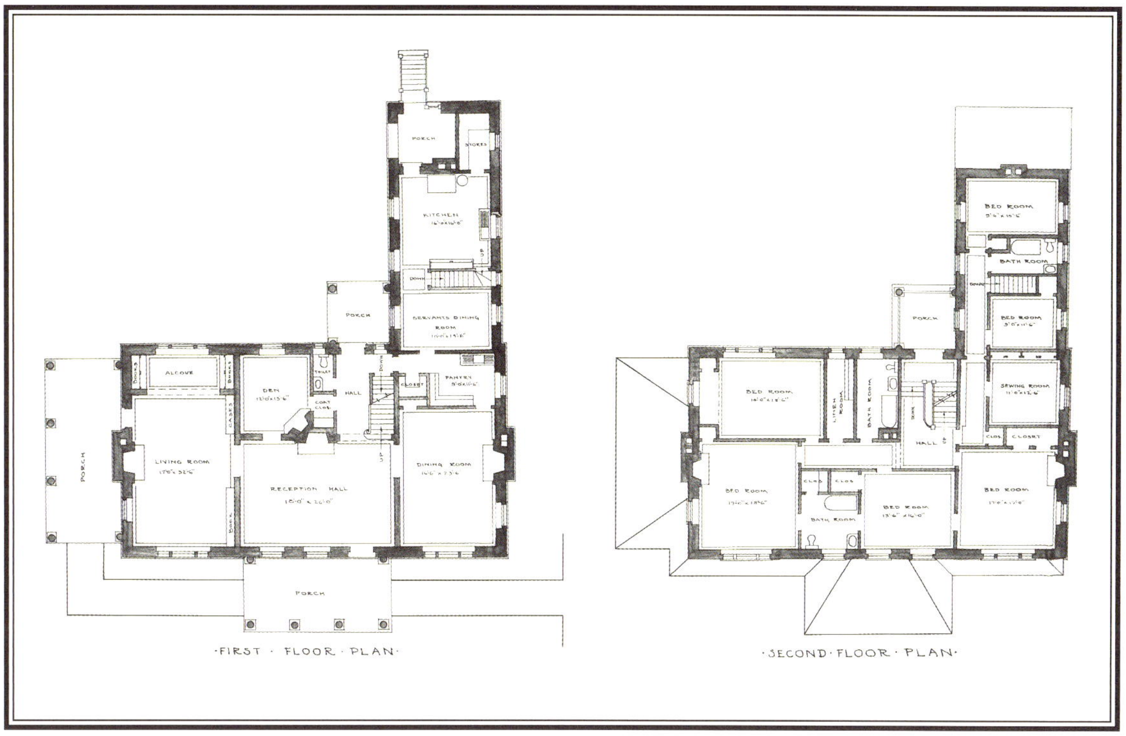

Above left and opposite: Bobby's new job required him to uproot his family of six and move to an area outside of Philadelphia. To accommodate the entire family, Beani found a 100-year-old, 22-room house in the beautiful suburb of Wyncote.

Above right: The floor plan boasted six bedrooms, four baths, a den, a reception hall, a formal living room, a formal dining room, a servants' dining room, a large kitchen, a pantry, a sewing room, a linen room, and four porches.

for the challenge."[6] Beani could not have been happier—she was going home.

That very weekend, Beani flew up to Philadelphia to purchase a new home. "I had to find a house in one weekend," Beani explained. "At the time, there weren't many five-bedroom homes there. Well, I found a 100-year-old, 22-room house. Bobby told me, 'You can have it, but I think you're crazy. You can't repair it. We'll live in it for five years, so you can get it out of your system.' And that's what we did."[7]

After the move, Bobby quickly settled into his new role as head of Chromalloy's building materials division, working out of the company's glass and aluminum corporate office in King of Prussia. Besides the increased responsibilities, Chromalloy offered Bobby a compensation package he could not refuse. In addition to a guaranteed base income of $35,000 a year (the same salary he earned while living and working in Florida), Bobby would earn a percentage of the profit from his father's original company, Architectural Glass Products.

Unrelated to Chromalloy, Bobby also received a commission on all mirrors sold in the state of Florida from Carolina Mirror. The management at Chromalloy became jealous of the added money Bobby made thanks to his business relationship with the mirror company. As a result, they offered him an increased salary of $75,000 to $80,000 annually, but only if he ended his commission arrangement with Carolina Mirror. Bobby explained:

> *Architectural Glass Products and Aldora Aluminum Products made more money than all of the other glass companies Chromalloy owned. The upper management really resented my bonus arrangement and other financial agreements.*
>
> *Upper management came to me, and they wanted to make a new deal. They wanted to double or triple my salary and then change my bonus arrangement. I told them, "I don't think that's fair, and I'm not going to accept it." It was like that the whole time I was up there. One year, we had a bad year,*

and I made almost no bonus. It was in single digits. It was almost nothing. They said, "See, we told you so," and I said, "I'm not complaining." They really didn't know how to handle me because big, public companies expect to tell everybody what to do. I'm a different individual. I said, "If you don't like it, fire me." They didn't do that either.[8]

Bobby told them that he liked the current arrangement and turned down the raise. The commissions and bonuses Bobby earned from Carolina Mirror resulted in annual earnings of $250,000 to $300,000, which was more than the corporate officers and executives at Chromalloy in St. Louis were making.

Loyal Employees

Bobby traveled extensively to all of Chromalloy's plants over the next five years. As a result, he met

and developed business relationships with a new group of industry insiders. Many of the people Bobby interacted with during the Chromalloy years now work at Arch Aluminum & Glass.

Mike Rogers, a longtime employee of Arch Aluminum & Glass, reflected upon his early days working at Chromalloy's Aldora division in Atlanta:

In 1971, I was just getting started, loading and unloading trucks. It was a summer job between classes. I was 18. Aldora, at that time, was very small. So, when the big guys would come through the local facility, everyone knew. That's how we got a chance to meet them. And, Aldora in Atlanta was growing. We saw sales double every month.

At the time, I was just a warehouse worker. I remember Bobby was a super nice guy. We all knew he was approachable. That meant a lot. It still does. Some of his peers in the corporate office, as I learned in later years, were nowhere near as approachable as Bobby was. I mean he's a good guy. Always has been.[9]

While working for Aldora, Rogers took architectural classes at night. Paul Copeland, the manager of Aldora Atlanta, saw promise in him because of his good work and ambition, and Rogers moved up the ladder quickly. "I had taken mechanical drawing at that point in time. Back in the old days, as part of our business, we used to do hand drawings," Rogers recalled. "We were doing what are referred to as shop drawings, drawings for the customer to see what they're going to buy so they can show it to an architect."[10] Rogers eventually moved into the aluminum side of the business.

Above right: This simple-looking truck, known as a boom truck, loaded and transported glass safely using a crane or boom. Prior to this advancement, workers manually loaded and unloaded the large sheets of glass.

Below: In 1971, the Silverstein family gathered for a picture to commemorate their eldest daughter Donna's bat mitzvah. From left to right: Donna, Cindy, Beani, Bobby, Ricky, the family dog, and Leon.

Don Beaver, who has had a lifelong career in the glass industry, met Bobby in 1972, when Chromalloy decided to enter the glass distribution business in Georgia. Bobby needed a knowledgeable person to run Architectural Glass, and he didn't need to look any further than Don Beaver. Beaver started at Chromalloy's Atlanta Architectural Glass division as the assistant manager under Paul Copeland.[11] But, in 1975, Beaver left Architectural Glass in Atlanta to manage a large competing company called Vega Glass. Bobby, unafraid of the competitor, told him that if things did not work out at his new job, he could come back to work at Aldora again. Beaver would take Bobby up on his offer and soon returned to Aldora.

Back at Architectural Glass, Beaver enjoyed working with Bobby, but he did not appreciate the financial demands the Chromalloy organization forced upon its managers. Beaver explained, "When it came to bonuses, if you made good money one year, you wouldn't make anything the next because Chromalloy would expect that much more of you."[12] As a result, the bonus program proved counterproductive to its recipients. It prompted Beaver to leave Architectural Glass in 1978 and start his own company—Thermo Glass Products. He began making insulating glass units and distributing glass, and once again, he was competing directly with Chromalloy. Around the same time, Paul Copeland also left Chromalloy for similar reasons. Likewise, he opened his own company and named it Tech-a-lite. Both Tech-a-lite and Thermo Glass Products would later merge with Arch Aluminum & Glass.

Another longtime colleague of Bobby's—Ray London, a cousin of Beani's—worked at Architectural Glass Products as the assistant general manager. Bobby explained: "After my father had died, I asked Ray, 'How would you like to come here to Miami and learn the glass business?' He decided to come down, and he moved near where we lived. Then, about a year later, I left to work for Chromalloy in Philadelphia. Ray stayed in Miami and worked for the new manager, who had replaced me. Well, I guess the new manager didn't like Raymond around because he's my cousin. So, when Chromalloy decided to open a place in Texas, the manager shuffled Ray off real quick. That's how Ray got to Texas."[13]

The Dallas plant opened in 1975. Like Beaver and Bobby, London struggled to work effectively in the chaotic, mismanaged environment created by Chromalloy leadership. Ray explained:

In 1973, just before the Silversteins moved back to the Philadelphia area, they celebrated Leon's 13th birthday with a bar mitzvah. Pictured left to right: Donna, Bobby, Beani, Leon, and Cindy.

> *The guys from Chromalloy would regularly complain about your inventory being too high. If you have a door, for example, and the door has five or six pieces ... you can't cut out one of the pieces to reduce inventory, because then you don't have a complete door. Those guys never understood that concept. So they'd make you cut your inventory, and your sales would go down. And they would complain about that. Then, you would increase your inventory, and you wouldn't be able to collect the money because everybody's order was back-ordered.*
>
> *Before Chromalloy took over, Bobby would tell everyone, "Run it like it's yours. You make good decisions, you make bad decisions—you won't have any problems that way." The guys from Chromalloy tried to integrate our glass and aluminum business in with its other glass and aluminum companies, and there was a clash of cultures.*[14]

Mounting Frustration

Eventually London would move on, and by this time Bobby, too, became increasingly discour-

When Chromalloy bought Architectural Glass Products and Aldora Aluminum, a total of three plants existed. As this catalog portrays, the plants were located in Miami, Atlanta, and New Jersey.

aged with the corporate hierarchy at Chromalloy. He believed upper management treated its employees poorly, and he felt stifled in the formal corporate culture. Still, in spite of Bobby's internal turmoil, the Architectural Glass Products and Aldora Aluminum Products locations thrived under his direction and brought in more money than any of the other plants in his division.

As a result of his frustration, Bobby openly defied Chromalloy's upper management team. Bobby explained, "One manager brought in a woman that he gave some authority to, and she obviously didn't know anything. I told the manager about the problems I had with her, but he wouldn't do anything about it. So, after several discussions with him, I moved out. I asked some of the men to come in from Pennsauken, the closest local branch to Chromalloy, and move my desk and all my papers and everything, and I set up an office there. On Monday morning, the manager said, 'Where is Bobby?' And the people in the corporate office said, 'We think he moved out. His desk is not here. Nothing of his is here. We think he went to Pennsauken.' So, I'm in Pennsauken and I'm doing the same job, but I'm doing it out of there. My boss said, 'What are you doing?' And I said I couldn't relate to the people there, and I needed to have my freedom in order to do my job."[15]

Bobby relayed another story about his rebellious nature: "Chromalloy had a standard company car, but when I went up there, I was driving a Mercedes. I almost always had a Mercedes. They didn't want me driving it, and I said I'd park it at the end of the parking lot. I gave them the kind of remarks I wouldn't tolerate from people who worked with me, but I wouldn't really treat somebody the way they treated me either. I'd just keep everybody happy."[16]

Beani elaborated on those troubling times: "Bobby didn't like the big business mentality, and he didn't like having people making his projections or changing the projections he gave them for budget purposes. He knew the business inside and out. He was born into it, and here he's going to have to run it on a corporate level with someone over his head. It didn't work. It just didn't work for him."[17]

Risking Everything

With continued frustrations at work, Bobby came home early on Halloween to take his son Ricky, then age nine, trick-or-treating. With his briefcase in hand, he took one look at his wife and said, "Trick or treat." After 20 years of marriage, Beani understood his hidden meaning. She explained, "I said, 'Oh, no! I don't believe you!' I knew what he did. He'd quit his job."[18]

Beani was shocked and dismayed by Bobby's impulsive decision. He attempted to explain to her how the management hierarchy tried to cheat him out of his commission deal as well as relay his other frustrations. Beani explained:

> *I made him go ask for his job. He didn't want to, but I was very angry with him for not discussing this with me because that was wrong. He said he*

couldn't describe how bad the situation was. Actually, they were reneging on a lot of their promises, and he couldn't convey everything to me. So he said, "Let's go and I'll show you." And he did show me. We went to his boss' house to talk to him, but he just wouldn't listen. Bobby was getting too expensive for him. His boss made it clear that he didn't want Bobby to achieve. They wanted him to do things below his abilities. I saw that Bobby was right. I knew it was never going to work for him there.

Bobby said it was time for him to do his own thing, and I agreed with him. We talked about what we had to do. First, we had to sit our family down and tell them. Bobby did not want to tell them that our life would change, but it would. We were living in this 22-room mansion. I loved to shop. I bought Cindy a new blouse every day. Life was going to change. Life was really going to change. We really had no savings. But, thankfully, the kids all understood. Donna, in her third year at Lehigh University, applied for and received a scholarship and a job. Leon had just gotten into Northwestern. We called up there, and the school said not to worry. Leon would also receive financial aid and a job at school. Cindy was fine with the situation, and Ricky was fine with it, too.[19]

Before heading off to college and with money tight, Leon worked for Beani's brother-in-law, Arnold Kramer. Kramer married Beani's sister, Mickey, and owned a chain of tire stores. Kramer explained:

Even as a young worker Leon was very energetic, very confident. He had a very good business sense. I put him in one of my stores to work, and he would get very upset if the manager did things wrong or if there was a petty theft. He was very sharp on every angle of the business.[20]

Word travels quickly in a specialized industry like aluminum and glass, and Bobby soon received a call from Max Korff, Bobby's cousin, who owned a glass company in Chicago and Boston called Solomon Glass. He asked Bobby to work for him, and with few other options, Bobby agreed. For the next eight months, Bobby traveled between Chicago, Boston, and Philadelphia. To make life a bit easier on Bobby, Max rented an efficiency apartment for Bobby in Boston.

During this time, Bobby began to formulate a game plan for the future. "Beani didn't like my traveling. I wasn't home much. I still couldn't afford a house. I wanted to go back to Miami," Bobby recalled. "I knew Beani wanted to stay in Philly, but I told her, 'Beani, the only shot I have is Miami. While I was up in Philadelphia for five years, I was an executive. I didn't know any of the people. When I was in Miami, I knew everybody. I helped [put] these guys in business and [kept] them there. I extended them credit and helped them set up their own businesses. That means something, or, at least it should.' "[21]

It would be three or four months before Bobby and Beani finalized their strategy to launch a business in Florida. The couple established a plan to visit several longtime glass and aluminum suppliers to rekindle business relationships prior to starting their new venture. That trip, they anticipated, would take several weeks, and send them down the East Coast to Florida. They hoped the meetings and resulting business connections would form the foundation for their future business pursuit. Three more months passed before the Silversteins would embark on their next business adventure: Arch Aluminum & Glass.

While out to dinner with Bobby, Beani, known for her artistic talents, doodled the design for the new company on her napkin.

CHAPTER FOUR

MOVE OVER, MIAMI

1978–1988

I'm not the smartest guy, but I am the most honest guy you will deal with.

—Bobby Silverstein

STARTING IN THE LATE 1970s, interest rates began to climb until they reached all-time highs of 18 percent and more. These exorbitant rates made it difficult for small companies to survive and grow. Bobby planned to open his business during this difficult period by targeting small local shops and high-risk credit customers he had serviced years before. Many of these smaller businesses would remain loyal to Arch from the beginning because, years earlier, Arch extended credit to them when other businesses would not.

Putting the Plan into Action

Mapping out the journey to reconnect with past contacts, Bobby decided to focus initially on procuring suppliers. To that end, Beani and Bobby took a three-week trip to meet face-to-face with suppliers Bobby had known for 20 years. Their goal was to establish a line of credit with these suppliers so Bobby could start his business. They traveled south, stopping first in North Carolina at Carolina Mirror. Architectural Glass Products had distributed Carolina Mirror products since the late 1950s, and that business relationship continued throughout the Chromalloy years.

Bobby believed his longtime alliance with its management team would prompt Carolina Mirror to do business with him now. To his dismay, Carolina Mirror turned him down and flatly refused to supply mirrors to his fledgling business. "They were still selling to my old company, and they were making big bucks. It was shocking because of our personal relationship, not because of the business decision," Bobby said.[1]

From there, he went to a glass company called American Float Glass in Kingsport, Tennessee. "My father was very close to this glass company. The management there said they would sell to me," Bobby explained.[2] Traveling farther south, Bobby trekked to Newnan, Georgia, and visited the William L. Bonnell Company. "They were an aluminum extruder. John Dunn was the manager's name. Aldora was a big account of Bonnell's. I went to speak with John, and we related very well. He said, 'Bobby, I'll do anything for you. Anything you want. And don't worry about paying for it.' So, I knew two things: I could get glass and [I could get] aluminum," Bobby remembered.[3] With these two items readily available to him, Bobby felt confident that he could have a viable business.

Bobby Silverstein started a successful glass and aluminum business with little more than his integrity and sterling reputation in the industry.

In this close-knit industry, Chromalloy quickly learned of Bobby's plans and took immediate action to thwart his undertaking. According to Bobby: "The head of Chromalloy's glass and metal division approached John Dunn at Bonnell and told him, 'If you sell to Bobby, we're not going to buy from you.' They continued to threaten to take their business elsewhere. Dunn did not worry about losing Chromalloy as an account. In fact, Dunn told the Chromalloy executive to find another supplier."[4] To this day, Arch Aluminum & Glass maintains a significant relationship with Bonnell.

Having secured suppliers, Bobby turned his attention to seeking financing. One relative offered to finance the business. However, this person proposed that Bobby would own only 20 percent of the company, and the investor would own the rest, along with maintaining complete control of the business. Bobby rejected the offer because he refused to let anyone else run the glass and aluminum business for him again. He intended to be his own boss at any cost.

Marvin Cravetz, one of Bobby's longtime friends, encouraged him to return to the Miami area and launch a business there.

Above: Marvin Cravetz, who had a dental practice, financially backed Bobby's business startup along with Bob Alpern.

Below left: John Dunn, on behalf of the William L. Bonnell Company, graciously welcomed Bobby back as a customer—despite Chromalloy's intervention—as evidenced by this letter of confirmation.

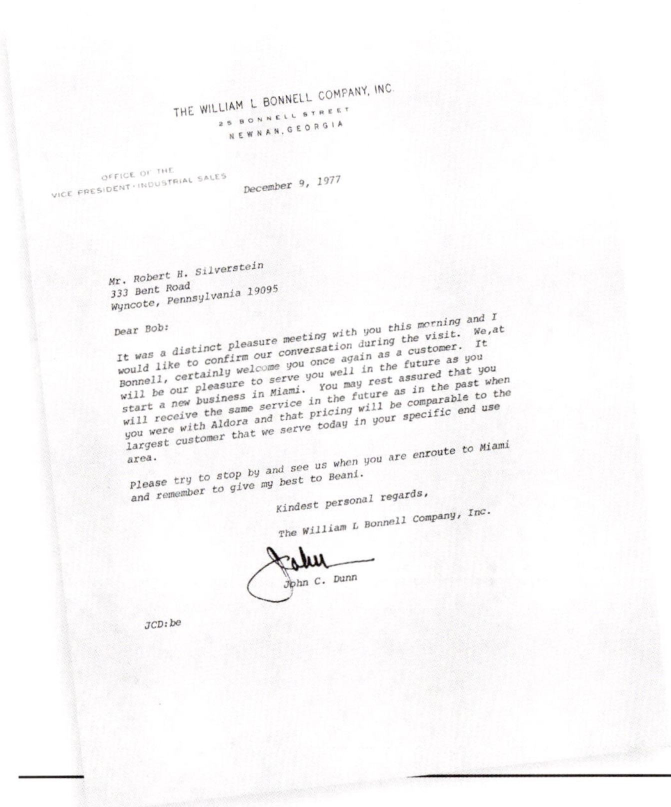

THE WILLIAM L BONNELL COMPANY, INC.
25 BONNELL STREET
NEWNAN, GEORGIA

OFFICE OF THE
VICE PRESIDENT · INDUSTRIAL SALES

December 9, 1977

Mr. Robert H. Silverstein
333 Bent Road
Wyncote, Pennsylvania 19095

Dear Bob:

It was a distinct pleasure meeting with you this morning and I would like to confirm our conversation during the visit. We,at Bonnell, certainly welcome you once again as a customer. It will be our pleasure to serve you well in the future as you start a new business in Miami. You may rest assured that you will receive the same service in the future as in the past when you were with Aldora and that pricing will be comparable to the largest customer that we serve today in your specific end use area.

Please try to stop by and see us when you are enroute to Miami and remember to give my best to Beani.

Kindest personal regards,

The William L Bonnell Company, Inc.

John C. Dunn

JCD:be

After Bobby encountered difficulties in raising capital, Cravetz, along with a glass salesman named Bob Alpern, offered to invest in his business. Confident in Bobby's ability to succeed, Cravetz and Alpern each contributed $75,000, while Cravetz put up an additional $75,000 as Bobby's share. Bobby gratefully accepted Cravetz's offer, and the three men went into business together.

Strapped for Money

With the investors' money in hand, Bobby sought a location for his startup business. He found a warehouse he could afford in a rough neighborhood in Miami, but finding a place for his family to live proved more challenging. A lack of money meant purchasing a home was out of the question.

Bobby did not believe in using investors' money for personal reasons, so he turned to another longtime friend, Alan Moses, for some

Bob Alpern, who worked as an independent glass salesman before joining Arch, provided Bobby with financial assistance to start the business.

immediate financial assistance. Years ago, Bobby and Beani had opened their home to Moses when he needed help. Now, he returned the favor and loaned Bobby money, no questions asked. Years later, Moses commented:

> *Bobby was always a very good friend. I could always count on him, and he could always count on me. We always had a great relationship.*[5]

The money from Moses bought a small three-bedroom apartment. Beani used her artistic skills to make their little home feel attractive and comfortable by sewing curtains and adding inexpensive decorative touches. Bobby sent his wife a small green plant as a housewarming gift and a thank you for making their apartment feel warm and welcoming. The little plant was the only gift he could afford.

The Final Details

Bobby's new company needed a name. He originally considered the simple "A Door and Glass Company, Inc." After discussing the matter over dinner, Beani suggested the name Arch. A-R-C-H were the first four letters of Sydney's original business, Architectural Glass Products, and many doorways had arches, Beani reasoned. She then doodled a logo on her napkin. Even today, her design remains the symbol and official logo of Arch Aluminum & Glass.

Having established the name, the financing, and the suppliers, Bobby moved on to setting up the books. He sought advice from Douglas Cove, an accountant who was Bobby's old golf partner from his days with Architectural Glass Products in Miami. Cove recalled: "In those days, you probably could have done the accounting in about half an hour on a single piece of paper ... a projection of what he was going to do for the first year, for the first two ... whatever he worked out, he didn't conceive of anything like this. You know, he did hit the numbers for his first-year projections. I think one or two customers made it for him."[6] Bobby had deliberately calculated his projections on the conservative side, and every year he would exceed his projections.

Back in Business

Bobby had the keen business sense, but he didn't have the mechanical know-how to build

How Arch Got Its Name

BEANI DECIDED TO ACCOMPANY her husband on one of Bobby's business trips back to Boston, and Bobby discussed his new business plans with her. Beani recalled:

> *I started getting excited about [them]. Then Bobby asked me, "What should we call it?" Suddenly it occurred to me—doors, entries, and porticoes. I was thinking about doors. So I said, "An arch! Oh my God, Bobby, this is so good! Arch comes before Architectural [in the dictionary and in the phone book], and that's the name of the old company." Bobby said, "That's dirty pool." I told him, "You've got that right."*[1]

Arch Aluminum & Glass opened for business in 1978. Arch rented this building at 395 North East 59th Street in Miami, which occupied 15,000 square feet of space. As Arch's business expanded, the company continually outgrew its locations.

storefront doors and frames or perform other functions necessary to establish his business. For that, he turned to his old friend and colleague, Arnie Studenski. Studenski still worked at Aldora, where he had worked years before when Bobby and Sydney were there. Bobby explained:

> *I asked him if he wanted to come to work for me. I told him I'd do something for him. I'd do whatever he wanted because I needed a handy person. Arnie said, "Let me think about it." I told him to take as much time as he needed. He came back to me and said, "You know, they haven't given me a raise in a while, and I don't really get along with the manager. I'm going to work for you." Then he said, "I don't even care if you can't pay me." Once the manager of Aldora heard Arnie was going to work for me, the manager told Arnie he would give him whatever he wanted in order to stay, but Arnie said no.*[7]

With a 15,000-square-foot warehouse in a rough neighborhood, one truck, and a handful of employees, Arch Aluminum & Glass opened for business on February 1, 1978. Bobby intended to reestablish business ties with customers he had previously serviced at Architectural Glass Products and Aldora, but he planned on doing so without undercutting his competitors' prices or stealing customers. Instead, he sought to win them back by providing the best service possible. Bobby explained:

> *When I started, I only went out in a 30-mile radius. I kept circling in the same area because I wanted to give quality service to the people. They all knew me, and I didn't want to screw it up. They said to me, "Bobby, you did so well for us." But I told them, "I don't want you to buy from me. Buy from the others. Only promise me that when you need somebody else to buy from, you'll give me the first shot." I knew I would hear from them. We started getting calls and doing business. We went farther and farther beyond that initial 30-mile radius.*[8]

After six months in business, Bobby received a call from John Dunn, the manager from Bonnell who supplied him with aluminum extrusions on credit. Dunn said, "Bobby, when do you think you're going to start paying me?" Bobby answered, "John, I was just waiting for you to call me. I'll start paying you soon."[9] And he honored his word.

A Father and Son Team

Even as a toddler, Bobby's oldest son Leon felt comfortable at work with his dad and his grandfather. During his college years, Leon spent every summer working in the warehouse at Arch, though Bobby never encouraged Leon to join the family business. But Leon enjoyed working there. Toward the end of one summer break from college, Leon told Bobby that he wanted to take a semester off and work at Arch. Bobby's response was clear: "I said, 'Leon, you should go back and get your diploma, but it is your decision.' "[10]

Leon decided to take the semester off and started in the back at the warehouse. He learned the business working his way toward the front office. He did everything from driving trucks to delivering merchandise. His childhood experiences playing in the warehouse, riding with the deliverymen, and visiting with the workmen helped Leon feel at ease working in the business as an adult. He felt comfortable in the blue-collar environment. After the semester ended, Leon returned to college, changed his major from industrial engineering to communications, and later graduated from Northwestern.

Upon graduation, Leon's future in-laws offered him the opportunity to talk about a job in their successful family business, Goldenberg Candy Company, in Philadelphia. Despite the generous offer, Leon elected to work at Arch with his father.

First, Leon worked two years as a truck driver for Arch. Then he spent five years in the factory earning his way to plant manager. Finally, Leon shared an office with Bobby. Bobby taught Leon the workings of

KEEPING PROMISES

BOBBY KNEW HE COULD NOT MANAGE the mechanics of setting up a glass and aluminum plant on his own. He also knew that to do the job right he needed Arnie Studenski to help him. Bobby boasts that Arnie has the skills to put anything together. He asked Arnie to come and work for him, and, without too much thought, Arnie agreed to leave Chromalloy and join Bobby in his new business. To show Arnie his appreciation, Bobby promised to buy him a new car. Arnie recalled:

> *One day, after I started working for Bobby he said, "Do you want to go to lunch with me?" I said, "Okay," and we went to lunch. After lunch, Bobby stopped at a car dealer, and he bought me a brand-new car. It was a Cadillac. He stopped and bought me a new car! That was a surprise.*[1]

Arnie Studenski has a long history with Bobby dating back to the days of Architectural Glass Products. Bobby credits Arnie's mechanical know-how for the successful launch of Arch.

the office, just as Sydney had taught him. Bobby allowed Leon to join in meetings with vendors and employees and participate in all the aspects of the business, but, echoing the words of his father, he told Leon: "Sit in my office. Sit at this desk. You will sit there, you will hear, and you will listen. If you don't agree, just write it down. If you have something to say, say nothing. We will talk about it after."[11]

Leon and Bobby, though possessing individual management styles, worked well together. The business sense of father and son complemented each other and served as an asset for Arch. "Bobby is a very bright guy. He gets to the nitty-gritty of things quickly and doesn't waste a lot of time on stuff that could be wasteful," said Douglas Cove, Bobby's original accountant. "He's very [focused] and knows where he's headed. I guess the impetus came from his son to go even further. So the combination of the two of them just clicked. It made the company grow."[12]

Leon took on management responsibilities and a new approach to fostering growth. He felt Arch

Rescue Mission

With the car jam-packed with their belongings, the Silverstein family started the long journey down the East Coast to Florida, where they planned to begin a new chapter in their lives. The trip went smoothly until they reached Fort Pierce.

Suddenly, the car conked out and could not be repaired. The family was stranded. Bobby had friends at the Aldora plant in Miami whom he hoped would help them out. He called his old colleague, Roger Sheets, and Sheets relayed Bobby's predicament to his coworker, Roger Collette. Collette, who loved to drive, willingly got into his car and set off for Fort Pierce.

During the ride back, Collette learned of Bobby's plans to open his own glass and aluminum business. Six months later, Collette went to work with Bobby, followed shortly by Sheets.

That ride would be one of many trips Collette would make over the years to help out Bobby. For example, early in Arch's history, Bobby got a call from a contractor in need of special glass. The contractor was building a racquetball club, but could not proceed with construction until he had the special glass in hand. Bobby asked Roger to drive immediately to Toledo, Ohio, to pick up four cases of heavy tempered glass, and then drive those cases down to Miami. To help ensure a successful trip, Bobby bought Roger a new set of tires for his truck. Enjoying the ride and the new set of tires, Collette believed he had received a great deal.[1]

From the early days at Aldora, Roger Collette never refused a favor for his friend and employer, Bobby. He would assist Bobby in business both on and off the road.

The Tech-a-lite operation, originally owned by Paul Copeland, was housed in this building on the north side of Atlanta.

should have a greater presence across the country. Together, the father-son pair led Arch Aluminum & Glass to flourish during the 1980s, with the opening of new facilities in and out of the state.

In three short years, Arch Aluminum & Glass outgrew its modest facilities. The Silversteins decided to move the business out of Miami and rented a facility in Fort Lauderdale. This new plant occupied 30,000 square feet—double the size of Arch's first plant. Soon after, Arch would need to rent 15,000 more square feet to accommodate its growing business.

Acquisitions Begin

While Arch Aluminum & Glass considered expanding, both Thermo Glass Products (owned by Don Beaver) and Tech-a-lite (owned by Paul Copeland) were thriving in Atlanta. "Everybody thought Arch owned Thermo Glass Products and Tech-a-lite, and the reality is, we didn't," Leon explained. "There was a handshake agreement that allowed both Tech-a-lite and Thermo Glass to use the Arch name and literature to sell product. We used this method to build brand equity."[13]

Bobby and Leon saw great potential for a glass and aluminum facility in the Atlanta area. After Paul Copeland died in 1982, Arch purchased the assets of Tech-a-lite from Copeland's estate. It then obtained majority ownership of Thermo Glass Products from Don Beaver. Both facilities remained in operation as they had before, but now they were known as Arch Aluminum & Glass of Georgia.[14]

Arch was continuing to grow when, in 1983, a customer who owed Bobby a good deal of money refused to pay his bill. In lieu of payment, Bobby and Leon agreed to take possession of the customer's real estate holding—a building in Fort Myers. Always alert to market demands and the growth possibilities of customers on Florida's west coast, Bobby and Leon decided to use the small building as their first

distribution center annex. To this day, the annex remains in business as a one-man operation.

With the addition of this annex, customers on the west coast could come to the facility to pick up purchased materials. This accommodation benefited customers who ran small operations and wanted their products quickly. They could pick up their purchases the next day, rather than waiting a week or more for delivery.

Recognizing the advantages, Arch quickly opened another small distribution center in Sarasota, Florida. With these two facilities in place and with its main plant in Fort Lauderdale, Arch could readily supply aluminum and glass to the entire southwest and central areas of Florida.

Above and below: For many years, Thermo Glass Products and Tech-a-lite conducted business in separate locations. After Arch bought both facilities in 1982, the businesses were combined and moved to a newly constructed plant in Austell, Georgia. This plant tempered glass as well as fabricated aluminum.

Opposite: These pictures depict an addition built in 1989 to make room for the tempering furnace in Austell.

Customers, particularly the smaller shops, appreciated the quick service time. Catering to the smaller companies became a target market for Arch, as it is today. Richard Davis, a customer of Arch's for more than 30 years who owned a small glass shop in North Carolina, described his experiences with Arch back in the 1970s and 1980s. Davis recalled:

> *Years ago, it was hard to get delivery on any kind of architectural metals or flat glass other than by ordering large truckload quantities. We used to buy it that way because it was the only way to get it to and from your jobs.*
>
> *That is, until Arch started delivering up in this area. They delivered once a week from the Atlanta plant, and we would get orders up to them. Of course, now they have a plant in Youngsville, which is pretty close by to us, and now they give most customers every day delivery. They pretty much changed the way the glass business is done in this area. Also, they maintain quality sales people who are very knowledgeable ... doing the technical work.*[15]

To better cater to customers and recruit workers in the Atlanta area, Arch saw the need to con-

solidate Tech-a-lite and Thermo Glass Products. "The Tech-a-lite operation was on the north side of Atlanta. Thermo Glass ... was located outside of Atlanta. These facilities didn't have the capacity for future growth in fabricating glass and metal," Beaver explained. "So, in 1987, we built a new facility in Austell, Georgia, which is just west of Atlanta."[16] The opening of this new facility, which was larger than the other two previous plants, allowed Arch to combine Tech-a-lite and Thermo Glass, and thus centralized operations.

With the addition of the Tech-a-lite and Thermo Glass facility, six operations now shared the Arch name: two in Atlanta, two on the west coast of Florida, one in Fort Lauderdale, and one in Texas (after leaving Chromalloy back in 1980, Ray London had set up an aluminum fabricating plant called Arch Texas in Dallas, and Bobby allowed London to use the Arch name, its inventory, and its catalog).

Moving Up

In 1986, just three years after the move to Fort Lauderdale, Arch once again outgrew its main facility and moved to Pompano Beach. There, Arch originally occupied 65,000 square feet. In just 10 years, this facility needed to expand first to 90,000 square feet and finally to 100,000 square feet to accommodate the company's growth.

The company grew rapidly—faster than anyone had envisioned. But not everyone wanted rapid expansion. Bob Alpern, one of Arch's initial investors, felt uneasy about the progress. Looking more toward retirement than expansion, Alpern showed less interest in the business. One day Bobby followed Alpern, only to find him at flying school. Bobby, fed up and frustrated with his lack of concern for the company, put his business card on Alpern's car to get his message across. Eager to spend more time flying planes and playing tennis, Alpern agreed to a buyout by Bobby and Marvin Cravetz. He would receive more than $500,000 in payment. Everyone was satisfied.

A year later, in 1988, Beaver sold his remaining interest to Bobby and worked as a consultant for Arch until 1995. Simultaneously, Arch opened a new facility in Pennsauken, New Jersey. This facility fabricated architectural aluminum and distributed glass.

BUILDING LOYALTY

IT IS EASY TO UNDERSTAND WHY ARCH boasts a loyal base of customers and employees. A sense of caring shapes Arch's relationships with its clients and with its employees. Ray London, who has worked with Bobby since the early 1970s, articulated his reasons for Arch's success:

> *I think our customers are very important. We treat them like they're more than customers actually. We spend a lot of money on service. We have very good products. The people who work for us are probably better than most people who work for other companies. They have a stake in the company, though it's not financial. Not stock or anything like that, but they have a stake in the company because as the company grows, they grow, and as the company does better, they do better.*
>
> *Our philosophy, Bobby's and mine, has always been to make sure we create a workplace that's safe, an environment that's safe and also stable for our employees. Never had a layoff. Hardly ever fire anyone.*[1]

The Big Time

In 1987, Arch Aluminum & Glass opened its first glass-tempering facility in Orlando. This opening proved strategic for Arch, enabling it to not only distribute glass but also fabricate it. "[Until this time] the only thing we owned, and the most expensive piece of equipment we ... owned, was the forklift. The trucks, we leased," Leon recalled. "Then we bought our first major piece of manufacturing equipment, a tempering furnace."[17] This pricey purchase moved Arch from a distributor to a manufacturer. The mom-and-pop shop had expanded its capabilities with the ability to temper its own glass.

Although the setup of operations at the Orlando plant appeared to run smoothly, Bobby and Leon soon learned of serious problems with the management team there. Bobby made a surprise visit to the plant to address the issue. His visit proved that he was not comfortable with the people working there. With only two weeks left until the opening of the plant, Bobby fired everyone except a skeleton crew. Despite this setback, the facility opened on schedule.

During the planning and opening of the Orlando operation, one of Leon's strongest business skills emerged. Thinking back upon his youth and his involvement with the glass business, Leon explained:

> *I would say I always tried to improve upon something I saw. I wanted to make things better. I would improve upon things, and most of what I improved upon was product flow, how [the glass or the aluminum] moved around the plant. I would see things.*
>
> *I anticipated problems with the design of the Orlando plant. I saw how the raw materials were coming into the building and how they were going out to be loaded. I changed the design so the flow was in the shape of a horseshoe. I wanted everything to flow to our loading doors, whether it was aluminum or glass. I wanted all the shipping and receiving to come to one place. Products came in one end and went across the back wall. To this day, every one of our plants almost looks the same.*[18]

This setup became "the footprint" and proved extremely successful. Adopting one of Bobby's basic business principles—"Don't change it if it isn't broken"—Leon would design the footprint for almost every facility.

Entering the Computer Age

As the company expanded, so did the need for advanced administrative systems. Bobby and Leon recognized that the business would benefit from expediting the process for tracking and processing orders and payments. The time for Arch to enter the computer age had arrived.

Bobby turned to his trusted son-in-law, Kevin Aubuchon—Donna's husband—for help. At the time, Donna and Kevin were living in California, where Donna worked at Apple Computer, having earned a degree in industrial engineering, and Kevin worked at ADP (Automatic Data Processing). "I did a little bit on the side for Arch … a little accounts receivable. I officially started working for Bobby in 1986," Kevin explained. "Donna was pregnant with our oldest. That's when we decided to move to Florida and for me to work at Arch."[19]

The timing was right for Kevin and Donna to make the big move east. They wanted to raise their children close to family, and the business needed Kevin's expertise. Bobby offered Kevin a job for $35,000 a year. "… Talk about riches to rags. We made over $100,000 a year in California and then we came here," Donna recalled.[20]

Initially, Kevin's duties involved computerizing the administrative offices. But, in order to develop computer programs for the entire business, Kevin needed a working knowledge of the company. So Bobby sent Kevin to the back of the warehouse to learn every aspect of the business. Kevin recalled:

> *It was difficult for me to go from the corporate environment to a family environment. That was difficult, especially being a son-in-law. You have rights, but believe me, you have to earn them.*[21]

Kevin Aubuchon, Bobby's son-in-law, worked on developing an efficient computer system for Arch Aluminum & Glass.

A Dog's Tale

As a result of her well-known fondness for animals, Natalie Zabric, a longtime Arch employee, earned the nickname "cat lady" from her coworkers. One day, while leaving work, she backed her car out of its space in the parking lot and accidentally ran over the Silversteins' beloved springer spaniel named Toby. Fearing for the dog's safety, Natalie leapt from her car and saw that the dog appeared lifeless under her vehicle. While crying uncontrollably, Natalie lamented, "I've killed Toby."

Several men from the warehouse came to her assistance and literally lifted Natalie's car off the ground to retrieve the dog. Surprisingly, Toby had merely been trapped beneath Natalie's car. To everyone's relief, Toby was uninjured, but Natalie was quite shaken up by the whole ordeal.

Above: Natalie Zabric at her desk petting Shadow, another dog owned by the Silversteins.

Right: Toby as an adorable puppy.

I worked in the back for years before I had a full understanding of the business, and then we went and found some small-end software and hardware for Bobby. At that time, we just did accounts receivable, invoicing, and inventory. The general ledger was still handled outside by Douglas Cove, Bobby's first real accountant.[22]

Kevin did not mind working in the back because of the comfortable culture established by Bobby. In fact, he looked forward to working with the Silversteins. "I was impressed by their work ethic, and I remember in the early days, working in the back and learning the business and trying to find an appropriate computer system and then implementing it," Kevin explained. "I'd work for three months in a row without having a weekend off. But it was nothing compared to how the rest of the family worked. You knew why they were successful. They were serious about what they did, but they had fun doing it."[23]

Kevin enjoyed working with Bobby. "Bobby was like a father to me. We were very close," he explained.[24]

Before moving to Florida, Kevin and Donna Aubuchon enjoyed their life in California but decided to move east in 1986 so they could raise their children near family.

Kevin also has deep admiration for Leon and his intuitive understanding of the business. "Leon has changed a lot over the years. We all have matured," Kevin recalled. "We were so young, not long out of college. Leon is a bright, bright guy. We would never be this size if it wasn't for Leon's aspirations, and Leon has learned to rely on people. To be successful, it is about the people around you."[25]

Through all the positive changes Arch saw in the 1980s, the homespun company culture remained constant. "We haven't subscribed to any slick new management methods touted in the bestsellers," Bobby explained. "We run our business so simply it would probably upset the people who write those books."[26] He refused to create a corporate structure that could interfere with a customer's ability to talk with a branch manager. Bobby's unpleasant experience at Chromalloy taught him that such barriers resulted in disgruntled employees and dissatisfied customers.

Arch ended its first decade in business by opening a second tempering plant at the Austell facility. With this addition, Arch completed its shift from solely distributing glass, mirrors, and aluminum doors to distributing, fabricating, and manufacturing—giving new life to Bobby's vision of the one-stop shop.

As Leon grew as a manager, his somewhat strong-willed and forthright manner mellowed. Leon developed and implemented a methodology for growing the company, and his vision would take clear shape in the 1990s. Leon's future challenges involved strategically expanding the company nationwide, introducing the company's unique philosophy, and opening new markets, while maintaining the much-cherished, family-owned business culture.

Arch produced this brochure, which showcases its product line, after Leon Silverstein retaliated against the owners of Sunshine Mirror by building an Arch Mirror facility directly across the street from the Sunshine plant.

CHAPTER FIVE

INNOVATION AND GROWTH

1988–1995

We were operating under the radar screen.

—Leon Silverstein

STARTING IN THE 1990s, FLOAT glass, which had accounted for 90 percent of all glass used in the building trades, was no longer sufficient to sustain a glass business. New uses for glass in architecture and new technologies greatly altered the glass industry. This decade saw the beginning of glass-tempering and an increase in laminated glass, which offered performance attributes such as improved safety. These changes started in the 1970s with the gasoline crisis, which increased the cost of production. The fabrication of specialized glass evolved during the late 1980s due to market demand, popular culture's interest in environmentally safe products, and the evolution of building code standards.

The federal government awarded grants to companies that developed environmentally friendly and more cost-effective technologies during this time. The country was in the midst of a serious recession, and all trades struggled to survive. According to the U.S. Department of Energy, the crisis of the 1970s had a profound influence on the glass industry. The high cost of converting glass plants to use energy-efficient fuels caused 15 plants to close between 1979 and 1983. Increased competition, environmental compliance, and labor shortages caused 33 more plants to close between 1982 and 1992. Because of the expense associated with shipping glass containers overseas, domestic markets were the only feasible customers. As a result, major plant consolidations occurred during the early 1990s.[1]

Arch's ability to stay one step ahead of the market influenced the company's growth and position in the industry. With Leon at Arch's helm, the business would evolve, expanding as a glass fabricator and moving toward greater investment in technology.

But technology alone wasn't what propelled Arch toward success. The leadership team at Arch recognized the importance of employee and customer satisfaction as well. Simply put, without its employees, Arch could not function. This appreciative attitude allowed Arch to maintain its family culture during this period of amazing growth.

Gerald Ashley started working at Arch in Orlando as a floor sweeper in 1987. He has moved his way up, working in a variety of positions. He currently serves as a branch manager at the Arch plant in Youngsville, North Carolina. Ashley explained what makes Arch so special:

Leon worked diligently to build Arch into a nationally recognized enterprise. He successfully expanded the business through his well-honed negotiation and networking skills.

Arch can take a person with no experience and train them in every area to become a successful branch manager. The company definitely strives [to promote] from within. And I'm a perfect example of that.

After I saw the opportunities and training given to me by my plant manager, I [built] an extraordinary team that enjoys coming to work every day. It creates an environment with less mistakes that provides a bigger and better bottom line and greater profit.[2]

After 20 years at Arch, Gabe Garcia knows firsthand the added value Arch provides its customers through its services. *(Photo courtesy of IDS Photography.)*

Another loyal employee, Gabe Garcia, joined Arch in 1985 as a college student who was studying architecture. He came to Arch to get experience as a draftsperson. Now, Garcia manages aluminum products for Arch's main branch in Tamarac, Florida. Here's his story:

At the time I was working as a cook. Arch offered to pay me what I was making as a cook, so I came on board for the experience, and I wound up staying. I did mechanical drawings, worked as a draftsman, and in engineering support.

My first impression was that Arch was really homey. It was and is a family-type atmosphere, and that's what I like most about it. The owner brings his dog in. It's not like really being at work and that feeling has not really changed over the past 20 years.[3]

The Growth of the Green Glass Industry

With the country clamoring for energy conservation and fuel efficiency, the glass industry responded. According to an article in *U.S. Glass* magazine called "Easy Being Green?" the American Institute of Architects and the International Union of Architects signed a joint Declaration of Interdependence for a Sustainable Future in 1993. The declaration led to a rating system for buildings. Simultaneously, the U.S. Green Building Council created the Leadership in Energy and Environmental Design (LEED) Green Building Rating System™, a voluntary national standard. These standards are based upon a point system arrived at by rating the categories of quality of water, energy efficiency, building materials, indoor environment, and design innovation.

Building designs now provide for more complex low-emittance (low-e) coatings—a microscopically thin film that helps to regulate the amount of solar heat absorbed and retained in the house—and the use of strategies to incorporate electrical savings through natural light. Studies found that business and educational institutions benefited from the natural environments because people felt a greater sense of comfort within "green" buildings.

The desire for sophisticated glass styles and treatments, such as glass louvers to control sunlight, self-cleaning glass, and laminated glass, as well as energy-efficient means of production, has radically altered the manufacturing and fabrication of glass.[1]

Another longtime employee, Neil Murphy, came to Arch in 1986 after working at Aldora Aluminum in Atlanta. "I've always found that Bobby was a man of his word," he explained. "He did what he told you he would do. You just could not work for a finer company than Arch or work for finer people than Bobby and Leon, the Silversteins."[4]

Managing the Customer

While Bobby and Leon take pride in helping glass and aluminum providers succeed, they frown upon customers who demonstrate disloyalty or negative attitudes toward Arch or its employees. Art Jackson, the plant manager at Arch Orlando, said, "Some customers believe they should be treated special and act disrespectful to the employees at Arch. Bobby has no problem telling them they will have to take their million-dollar account somewhere else."[5]

When it comes to customers, sometimes Bobby finds it necessary to confront those who do not pay in a timely fashion. In one instance, he decided to visit an Arch customer who refused to pay his bill. Bobby's longtime associate Roger Collette discouraged Bobby from taking the trip alone and offered to join him, but Bobby just laughed, told Collette not to worry, and headed out the door. He didn't know that the customer was a biker from California. Nor did he know the man had been released recently from prison.

He arrived at the customer's business and was shown into his office. "He was a big guy," Bobby recalled. While Bobby sat across from the customer with his usual deadpan expression, the man accused him of harassing his wife because Bobby had called her looking for payment—she was the bookkeeper.[6]

Suddenly, the customer walked around from his desk, positioned himself in front of Bobby, and assumed a karate position. Bobby looked at him calmly and said, "I don't know what you are trying to do. And if you are trying to antagonize me, there's nothing for me to do." Bobby later admitted he was the most scared he had ever been in his life, but, at the time, he remained absolutely still. The customer looked at Bobby and then returned to his seat behind the desk.[7]

The two sat in silence. Finally, the customer walked to the office door and opened it. "If I were you," he told Bobby, "I would get out of here right now." Bobby got up, walked over to the customer, and put out his hand. Automatically, the customer returned the handshake. It has always been Bobby's philosophy that the best way to catch someone off guard was to be nice to them. Flustered, the customer grew irate. He warned Bobby, "If I were you, I would get into my car real fast." Bobby calmly replied, "Thank you."[8]

Bobby quickly left. The customer declared Chapter 11 the following day, and Bobby never received the payment he was owed. According to Bobby, "The moral of the story is: Don't antagonize

By making the major investment to produce its own powder paint, Arch met the needs of its aluminum customers who wanted the superior finish that powder paint provided. Arch offered a wide variety of standard colors, which enabled the company to produce and deliver the product quickly.

What's in a Name?

BOBBY HAS TWO SISTERS—PAULA IS the eldest. Although everyone calls her Paula, her actual name is Helen. And so it only seems logical that, in school, Bobby was known by his middle name, Harris, despite Robert being his given name. One day, when Bobby was in kindergarten, Paula was sent to get him from school. She recalled:

> *He was five and I was almost nine. We called him Bobby at home, but when I went to the classroom door and I asked for my brother Bobby, they brought out another child. I told them, "That's not my brother. My brother is over there." They said, "No, that's not [Bobby], that's Harris." Nobody at home ever told me that my brother [went by the name] Harris. But, actually, Bobby had a lot of names. He was king of the family.*[1]

anyone. If you do, then they become the aggressor. People are always looking for a reason. Don't give them a reason."[9]

Taking Control

Leon has always believed that business size is equal to leverage, and with leverage comes power. In 1989, Arch, though in the process of expanding, was still viewed by the big suppliers in the industry—such as PPG, Pilkington, Guardian, and Ford—as a small, family-run glass business. Leon wanted more and worked hard to alter that perception. However, from his perspective, even with expansion, it remained imperative to maintain control of the entire process—from production, to delivery, to service.

One example of Leon's ability to take control of a process occurred in 1989. Around this time, U.S. lawn mower, lawn furniture, and playground equipment manufacturers embraced a painting process that used powder paint as the coating for metals. However, despite interest from customers, the aluminum industry, along with most architects in the United States, resisted changing from traditional high-performance wet paints to the powder-paint coating, due to the prohibitive costs involved.

The wet paint finish uses volatile organic compounds (VOCs), which are toxic and have been outlawed by some states. "Some high-performance finishes require a pretreatment of acid chromate, and that's heavy metal. You can't neutralize it or send it down the drain, you have to barrel it up and bury it somewhere," said Paul Parcell, a former self-employed glazier, who joined Arch in 1990.[10]

Arch, like other aluminum fabricators, sent its metal to an outside vendor for the wet paint process. By sending the product out-of-house, Leon reasoned that Arch lost control of its ability to provide prompt

Bobby recognized the importance of employee and customer satisfaction, but he frowned upon nonpaying customers. Often Bobby would personally call on customers who were behind on their payments.

delivery and service because the product was vulnerable to the vendor's schedules and demands. Therefore, Leon investigated the idea of bringing the painting process in-house using the more environmentally friendly paint process. He discovered that if Arch created a finishing line with the powder paint, the entire extrusion and paint process would be reduced from six weeks to just two, and Arch could control quality for the whole operation. Bringing the powder paint process in-house would reduce costs, speed delivery time, and provide environmental benefits as well. "Our powder paint process doesn't require acid chromate pretreatment, and our powder coating doesn't emit any VOCs," Parcell explained.[11]

"It takes capital to do these new things," Parcell continued. "When I came to the company in 1990, the whole industry was in a recession. Companies were failing. They didn't have the capital to put in new equipment. Even to this day, it's a big capital expenditure and a big way of changing the way you do business."[12] But spending the money was not a point of contention for Arch. According to Leon, "If we made money, we immediately invested it back into the business. We were never afraid of debt. We were always making money. Once we paid off one loan, we would invest in something else."[13]

Custom Software

To optimize production, become more efficient, and create the necessary infrastructure for further expansion, Arch needed order entry and manufacturing software. Despite the ongoing investment and involvement of the information technology team, Arch's employees still entered much data by hand; there was no software already available in the industry that fit the company's needs. The first software system installed at Arch cost approximately $10,000 and enabled employees to invoice, order, and track accounts receivable on a computer. However, the software did not provide services for manufacturing or fabrication.

To achieve full computer integration, Arch needed to create software specific to its needs, and Kevin Aubuchon would lead the team to do it. This customized software would help propel Arch to a $100 million business. "This was a remarkable endeavor considering we didn't throw a lot of resources at it," Aubuchon said. "No one else in the business has developed and installed their own systems."[14]

Few private business owners would enter into such a costly venture without hesitation, but this was and continues to be the Silversteins' standard approach to business. Garth Goodrich, a longtime glass supplier for Arch who has known Bobby and Leon since 1991, reflected on his 30 years in the industry:

> *Family-owned businesses typically are content to be what they are, what they inherit. But the Silversteins were more ambitious and reached out well beyond where they started. A lot of businesses either grow or go away. That's part of the world. But the unique thing is the way Bobby and Leon were willing to take chances and move well beyond their early start in Florida.*[15]

In 1988, Bobby's youngest son Ricky, who had just graduated from college, decided to follow in the footsteps of his older brother, father, and grandfather, and joined the family business. Unlike Leon, however, Ricky started his career at Arch in sales. Though he sat in on many meetings with his father and brother where major decisions were made, he did not pursue a prominent management position until years later. For the moment, Ricky happily worked from his office selling Arch's ever-expanding product line.

Leon's dog Max came to work every day and took his job of protecting Leon quite seriously. He had a soft side, though, and would come running when he heard The Birthday Song because he loved cake.

Mirror, Mirror on the Wall

Numa and Mary Handi, a husband-and-wife business team in the mirror industry, approached a salesman named Ed Hayes in 1989 with a proposal to manufacture mirrors in Florida. Hayes and Numa took the idea to Ron Brock, the president of Westshore Glass, and then they approached Bobby and Leon.

While managing Chromalloy, Bobby had met Numa and Mary in 1975 at Chromalloy's Texas mirror plant where they worked. Numa had almost 30 years experience in the mirror industry at that point. In 1984, he had decided to start his own mirror company in Texas, but, unfortunately, by 1988, Numa declared bankruptcy and closed his business.

Finding himself unemployed, Numa decided to follow a hunch. He contended that, contrary to an industry-held belief, a plant could successfully manufacture mirrors in Florida. Because of the extreme humidity and high salt content in the water, industry insiders argued that Floridian conditions made the manufacture of mirrors impossible.

This assumption arose based upon the experience of one mirror facility in Arizona. The plant utilized multiple water sources, and the municipality couldn't tell the company which was causing the problem. It turned out that some of the water was contaminated, and the poor quality made making mirrors unfeasible. The manufacturers had concluded that the high salt content of the water was to blame, and no one ever questioned the assumption. As a result, no one ventured to manufacture mirrors in Florida.

Numa would change all that. He convinced Arch and Westshore to consider forming a 50/50 partnership in a mirror company to be called Sunshine Mirror, in which Numa would set up the operation and run the plant.

But before Arch and Westshore would agree to the arrangement, they asked Numa to justify scientifically that conditions were appropriate to manufacture mirrors in Florida, and he agreed. Numa traveled to Fort Pierce, Florida, where the partners planned on building the plant. First, he tested the water quality and found that, after purification, the water was suitable for mirror manufacturing. He then conferred with the Fort Pierce water utilities district manager to ensure uninterrupted availability of water, because mirror manufacturing requires a steady flow of at least 70 gallons of water a minute.

With Numa's testing and research complete, Bobby, Leon, Ron Brock, and the owners of Westshore Glass looked forward to serving as the sole manufacturer of mirrors in Florida. They reasoned that both companies—Arch and Westshore—already held the key markets for distribution of mirrors in the state. While Westshore would prove less enthusiastic about future collaborations, at the time they supported Bobby and Leon's plan to serve as the administrators of Sunshine Mirror. Leon recalled:

> *Numa and Mary have histories with my father [that go] 30 to 35 years back, maybe longer. We were probably the largest distributors of mirrors in Florida, and Westshore Glass was probably the second-largest distributor of mirrors in Florida. Both of us had mirrors either coming from North Carolina or Japan. Japan was known for high-quality mirrors back then. We were [selling] enough mirrors that we could manufacture them internally, and neither of us thought we could do it ourselves. So we decided to do it as a joint venture.*[16]

Construction of the plant began immediately. As a result of Numa's years of experience in the mir-

Numa and Mary Handi were instrumental in the successful launch of Arch's mirror facilities. They have a long history with Bobby, dating back to 1975 when they all worked for Chromalloy.

Hurricane Prompts Improved Standards

THE DEVASTATION WAS HORRIBLE. There was $20 billion to $30 billion of destruction to homes, businesses, and schools. No one in its path was spared from the winds, pressure, and debris of Hurricane Andrew. According to the Florida Building Code Commission, research revealed that lack of enforcement, alterations to, and complete disregard for existing building codes contributed to the destruction.

At the time, Hurricane Andrew was the worst natural disaster the United States had seen in more than 30 years. In the aftermath, the nation's focus turned to natural disasters and improving building standards.

The attention prompted the industry to move toward creating and using impact glass, and improved and strengthened window and doorframe design and construction. After the hurricane, Florida led the country in setting building standards, although it took about six or seven years for the first cohesive guide to materialize. Those standards are revisited and revised every three years, and now local building codes must be compatible with state standards. Yet, interpretation of the codes still remains problematic.

A study of the hurricane's effects found windborne debris caused the loss of windows, doors, and roofing. And despite the hundreds of code-compliant products now available including doors, walls, and roofing, problems associated with windborne debris still exist. The search for ways of producing stronger glass continues.[1]

ror industry and his expert understanding of mirror production, Bobby and Leon helped out as necessary, but mostly they stayed out of his way.

Knowing that it would take time for the plant to get up and running and that Numa would have living expenses in the interim, Bobby wanted to pay him. While the owners of Westshore Glass did not agree, Bobby insisted on putting Numa on the payroll. "We will never forget that time," Mary said. "When you are down and out, people aren't too nice. Bobby and Leon are really good people."[17]

Numa designed and set up the mirror manufacturing plant, hired the staff, and initiated operations. While he worked on setting up the plant, Mary came to visit him and became an employee at the new Fort Pierce plant. Mary's extensive knowledge of the mirror industry enabled her to run all the administrative functions, including sales. Even with Mary's assistance, Numa required additional, knowledgeable employees to get the plant operational quickly. He suggested hiring the crew from his defunct Texas plant. Without hesitation, Bobby paid and financed eight men who had worked at Numa's Texas facility to work in Fort Pierce.

In 1990, after a long year of testing, building, and preparation, Sunshine Mirror opened its doors as a state-of-the-art mirror plant—the newest plant in the United States. The last mirror facility built in the United States was by that time at least 15 years old. All the other mirror plants still functioning in the United States had been in operation for generations.

Within two years of operation, business hit an all-time high. "We were also [a large] user of laminating glass, so we duplicated the mirror [model] for laminated glass," Leon said.[18] Arch and Westshore Glass established another 50/50 partnership, which they called Sunshine Laminating. Because the two companies sought a different customer base—Westshore Glass focused on residential construction, while Arch focused on storefronts and commercial glass—no conflict of interest existed in the joint venture.

Instead of moving to a different facility, the building that housed the mirror plant was expanded another 35,000 square feet for the laminating operations. Even though the mirror and laminating plants were under the same roof, they operated as two separate business entities.

Natural Disasters and Building Standards

Despite the seemingly prosperous times for Arch, the economy had taken a downward turn, while the building industry experienced a serious slump. Leon reasoned that diversifying and expanding the aluminum line would be beneficial for Arch, as business was slowing. Sunshine Laminating had opened in 1992, just months before Hurricane Andrew, a category-5 storm, struck the shores of South Florida. "We started the laminating business, but at the time, there wasn't a focus on hurricane windows," Leon said.[19]

Soon after the hurricane's devastation, Leon believed it was only a matter of time before the state of Florida implemented improved building standards and codes that would influence the manufacturing of aluminum doors and windows. Consumers already requested impact glass, even without building code requirements, and it would take several years to establish a uniform set of codes.

After Hurricane Andrew, Miami-Dade County, in partnership with the University of Texas, worked on creating a model that would test the ability of glass to withstand the impact and pressure of powerful winds.[20] Arch stood ready to respond, keeping a close watch on the progress and working on developing glass testing of its own. "We spent a lot of time and money investing in testing. We bought an air cannon to test the strength of glass," Leon explained.[21]

Arch became a key player in the testing and developing of Miami-Dade's building codes and standards. The effect of the hurricane also shed light on other related issues in the county.[22] "The manner in which so many neighborhoods had simply blown to pieces during Hurricane Andrew drew attention to the product approval system," Leon said.[23] The county was under pressure to come up with a model that could be used to set new and better standards.

Because Arch closely observed the results of testing by the University of Texas and engaged in its own testing, Arch became the first company certified by the county for its impact products. That relationship with Miami-Dade continues today, as

SUCCESS AT AUCTION

LEON SILVERSTEIN SEES A CLEAR rationale for his success at auctions. He explained:

> *The difference is [that] when I go to an auction, I'm the only CEO going. The others are sending managers or operations people, and I guess a lot of them are afraid to pull the trigger. The money in my pocket is my own, and the decisions I make are my own. Now, we will buy a lot of peripheral equipment, and our basic rule is if we can get it for 25 to 30 cents on the dollar, we'll buy it.*[1]

In 1989, the Silversteins hosted a party celebrating the purchase of a new tempering oven at the Austell, Georgia, facility. From left: Bobby and Beani Silverstein; longtime employee Don Beaver and Bobby's younger sister, Natalie Reed.

Breaking ground at the future site of Arch Mirrors. The Sunshine Mirror plant can be seen in the background.

does Arch's participation and commitment to creating safer national standards.

Learning the Auction Game

In the meantime, the Orlando plant supplied South Florida with large quantities of tempered glass on a daily basis. Bobby and Leon decided the demand for tempered glass justified opening another tempering facility, this time in South Florida. They chose to house the equipment at their headquarters in Pompano Beach.

Horizontal flat tempering ovens, which reduced energy consumption and produced a better tempered glass product, had replaced the outdated vertical hot-air tempering ovens, so Leon set out to buy the newer version. Because of the prohibitive price for new equipment (between $1.3 and $1.4 million), Leon was determined to purchase used equipment through an auction. Leon explained:

There was a company in Tampa, Florida, that had a furnace made by the same people who made the

Changing with the Times

Ron Brock, president of Westshore Glass and Sunshine Mirror, contends that the business of selling glass has changed radically over the years. He explained:

Back in the 1970s, you would sell cases of glass to people who installed glass. They would put the glass in the storefront and that was that. Today, everything is value added. Either it's energy efficient, or it's safety [glass], or it's hurricane resistant. All the glass that's being used is now a fabricated product. It is no longer just a raw glass product. So, as a distributor, we found that we no longer could just buy glass and sell it to the customers. We had to start fabricating as well.[1]

furnace we bought in Austell, Georgia. So we went to that auction. We were one of the smart people and had properly read about the auction procedures. You had to have certified funds available for your bid to be accepted. I bought the furnace for probably 15 cents on the dollar and found out that auctions were a very good place to buy good used equipment.[24]

With this purchase in hand, Arch could now easily supply the entire state with tempered glass.

Growing Pains

Up in Fort Pierce, Sunshine Mirror had reached capacity and needed to expand. Although Bobby and Leon wanted rapid growth, Ron Brock and the owners of Westshore were not interested in expansion outside of Florida. Nevertheless, Bobby and Leon started researching places to build a mirror plant targeting the Midwest.

At the same time, representatives from Iowa's economic development group visited Florida looking to attract businesses. The Iowa group approached Arch. Bobby and Leon expressed an interest in opening a plant in Iowa but did not have the financial wherewithal to make another large capital expenditure at the time. The Iowa group asked Bobby and Leon, "Would you be interested if you didn't have to put up any money?" For Leon and Bobby, the answer was easy—"Absolutely, yes!" With the deal in place, everything was arranged quickly, including Numa Handi's involvement in designing, building, and opening this new mirror plant.[25]

Leon, recognizing that Brock and the owners of Westshore did not want to grow the mirror franchise, decided that the present business arrangement with Westshore had to change. He approached them with an offer to buy them out of Sunshine Mirror. To Leon's disbelief, Brock and the owners declined.

Based on the original buy/sell agreement, Brock and the Westshore owners decided that they would buy out Arch instead. Leon was greatly surprised by this decision, but he was not thrown off balance. He replied, "If you buy us out, we are going across the street from Sunshine Mirror in Fort Pierce, and we will build a new mirror plant of our own."[26] Brock and the Westshore owners must have thought Leon was bluffing because they proceeded to buy Sunshine Mirror from Arch. Within weeks, construction began at Arch's new mirror plant across the street, with the help of Numa Handi.

The separation from Sunshine Mirror was complicated by the existence of the laminating facility and the joint ownership of the building. "We were technically still partners in the laminating plant. This time, instead of us offering them a buyout price for the laminating facility, we waited them out," Leon explained. "Finally, they offered to buy us out. At that time, they didn't think we had any money because we just spent a lot of money putting up two new mirror plants. But instead of us selling the laminating plant to them, we in turn bought it from them."[27]

Fortunately, for Leon and Bobby, their business relationship with aluminum manufacturer Aluminum Shapes had provided Arch with access to money through a bank allied with the company.

Arch was able to make the necessary counteroffer to buy out the laminating facility, and closed the deal within 24 hours.

Still, both Arch and Westshore had financial investments in the Sunshine Mirror building. The Westshore owners now leased 65,000 square feet for their mirror plant, and the Silversteins leased 35,000 square feet for their laminating facility. When the partnership was young, they had agreed to extend the physical space for the mirror line to cover 75,000 square feet, while only using about 25,000 square feet for the laminating facility.

With the change in business relationship, there were now two separate companies housed under the same roof. Leon, disgruntled with the arrangement, told the owners of Westshore:

> *Get out of our space. We have a lease. We want to build a wall between your mirror side and our laminating side. They said, "Well, you can't. We would have to shorten our mirror line to move the mirror plant out of the Sunshine laminating space."*
>
> *I was frustrated. One of their bridge cranes ran through the plant, and one weekend, I shut off the power. I was right there Monday when they came in to the plant, and they couldn't operate the cranes. They ran to court and cried foul. They accused me of messing them up. That did it. The judge heard the case immediately.*[28]

Leon got action from the court, which was the outcome he desired. "We went to court. The judge agreed with me," Leon said. "A lease is a lease. So they had to shorten their lineup."[29] But, instead of resolution, both parties experienced more frustration, more lawyers, and more court hearings until an agreement was reached.

Surprisingly, the owners of Westshore and the Silverstein family are still partners in the real estate. "We actually have a good relationship as partners in real estate, and whatever bad blood there was, time cured," Leon said.[30] The two companies were eventually able to coexist peacefully. "We were honest," Leon said. "We both wanted to make money in our respective markets. We fought about the building, but we did not take it to the marketplace."[31]

Opposite: When reflected on fabricated glass, the zebra-striped board can pick up distortions and other imperfections in the glass.

Below: At the Bettendorf, Iowa, mirror plant, workers carefully inspect the surface of a mirror for irregularities prior to shipment.

JUMBO GLASS

IN THE EARLY 1990s, A man came up with an idea to build a glass house. He wanted to use huge single sheets of glass—so large that each sheet would form a wall of the house. No one wanted to touch the job. "Call Arch," someone finally suggested. Arch had a reputation as a cutting-edge company that would take risks and take on challenging projects.

When the man called Leon, at first Leon did not take the request seriously. But he told the man that, in theory, Arch could make a piece of insulated glass 120 by 216 inches in size. However, it would require specialized equipment, and the cost would be prohibitive. According to Leon, the man said, "I just want one piece to try in my guesthouse. If it works, I want to build a glass house." Leon responded that even one piece of glass would cost a fortune. The man calmly answered, "Try me." Leon quoted him a price of $100,000 for the 120-by-216–inch piece of insulated glass, never thinking the man would agree to the high price tag. The next day the man wired a down payment of $50,000 to Arch with instructions and a note stating, "You're on!"

Arch agreed to hold on to the down payment with the understanding that it would take on the project, but, if it failed, Arch would return the money to the man. The agreement also specified that the man would need to visit the warehouse

and approve the glass before Arch attempted to ship it to his guesthouse.

The fabrication process began almost immediately. Because the glass would be so large and heavy, Arch rented specialized equipment, including a specific type of crane to lift and move it. Figuring out how to move the glass would be only one of the many quandaries Arch employees would face in regard to this project.

When working on the jumbo glass, the regular equipment needed to be moved to make room for the new equipment. Therefore, weekends were devoted to fabricating the jumbo glass, so that regular glass fabrication could occur uninterrupted during the week. To encourage employees to work weekends, Leon and Bobby offered workers a special bonus based on the customer's fee.

As workers solved one issue, new problems or questions arose: How can we ship such an enormous sheet of glass up a mountain road to its final destination in Montana? How can we install the giant piece of glass without breaking it? And, how can we fabricate the glass to withstand various temperatures? Working through each problem, Arch workers found viable solutions. At last, the glass was produced and ready to ship. As the trailer transporting the glass moved along the mountain road, the glass had to be lifted off the trailer by a crane at each curve in the road and then reset onto the truck.

Despite all the challenges, the glass arrived unharmed, and workers successfully installed it at the man's guesthouse. At the time of production, the glass was the largest sheet of insulated glass fabricated in the world. Pleased with the outcome and look of the glass, the man commissioned Arch to provide jumbo glass for three additional houses built over the next three years at a cost of more than $2 million worth of glass.[1]

Opposite page and below: By the mid-1990s, oversized glass panels had become very popular. Arch developed innovative ways to store, ship, and install the jumbo glass, resulting in an impressive finished product.

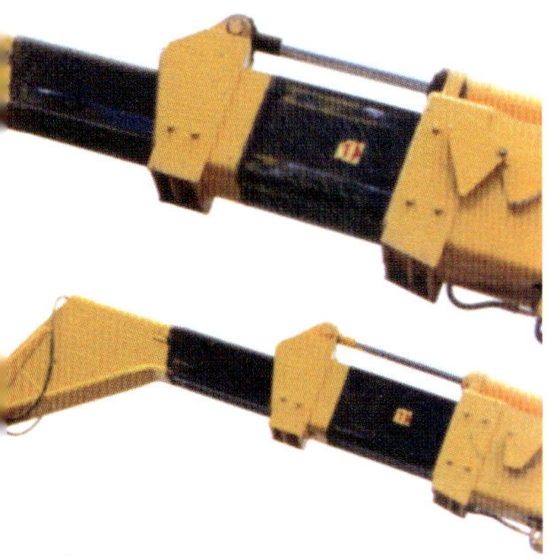

Building the Business

In the mid-1990s, much consolidation occurred within the glass industry. Arch needed size and leverage for purchasing power to put toward health-care, trucks, and other necessities. To meet this need and to fend off any threat of a buyout, the family created a working formula for expansion and planned to expand production in tempering and insulating glass. "We had an idea of how to run our type of business," Leon explained. "The goal was to get big fast because we didn't want to be bought out by somebody else, and we also knew if we didn't get to be a certain size, it would be tough for us."[32]

During 1994, Bobby and Leon made major business decisions to expand quickly. To determine where to launch their next few locations, they took several factors into consideration, including the population of the market, competitive conditions, and overall building activity in the area. "There was a method to the madness," Leon said.[33] Using a computer model based on these variables, the Silversteins would develop a list of potential sites. With this list in mind, they would often create market rumors about where they planned on building a new plant to help them decide between two places.

In the potential cities, they spread the word through customers and salespeople that Arch was considering opening a plant. Each area had its own rumor. Leon, using his keen business insight, expected that where the biggest buzz occurred would be the logical location to put the next plant. The noise emanating from the Indianapolis market was deafening, so Leon knew that an Indianapolis plant would work and set up a plant there.

At the same time, Dan Haney, the Arch Orlando plant manager, approached Bobby and Leon with an idea to build a business in Memphis, believing it would serve as a perfect extension to Arch's Atlanta facility. Bobby and Leon agreed, and they set up shop in Memphis. Their one fatal mistake concerned the selection of the actual location of the warehouse. They had not taken the time to investigate the neighborhood properly. Leon explained:

> *We picked a real bad warehouse in a real bad neighborhood, and we never did well there. The only month we ever had a shot at making money was the month that we found out that two of our employees were on the FBI's "10 Most Wanted" list. We called and arranged to have cops and a SWAT team come. But the employees got wind of it and escaped out of the back door. We didn't get the guys or the reward money.*
>
> *Because of the bad neighborhood, there was so much theft, so we kept hiring private detectives and putting them on our payroll. Basically, we'd hire them, and they'd be workers. You couldn't tell they were detectives. They'd work for two or three weeks just like everybody else, and they'd give us reports of everything going wrong out there, and the amazing thing was, everything went wrong. Everything was a problem. Everybody was bad. And, we had a five-year lease. We went through two or three managers until we finally just threw up the white flag.*[34]

The Silversteins learned a monumental and, ultimately, cost-effective lesson: future Arch plants needed to be located in first-class neighborhoods and in first-class buildings. Decisions on where to

After fabricating the largest piece of insulated glass in the world, Arch transported a replica of the glass to trade shows in recognition of this unusual accomplishment, and to tout the company's innovative ways.

WEATHERING THE STORM

IN 1985, MIKE ROGERS CALLED BOBBY Silverstein to inquire about a job at Arch's Fort Myers, Florida, facility. Rogers had known Bobby since their days at Chromalloy. Bobby flew him down to Florida from Atlanta to talk, and the meeting resulted in a job offer. Shortly after returning from the meeting, Rogers and his wife decided to visit Fort Myers on Labor Day weekend.

The couple booked a room at a motel outside of Fort Myers. Unfortunately, that weekend a hurricane remained stationary off the coast and drenched the area in rain. They ventured out in the pouring rain to head to the city that Sunday, but the drive took hours as a result of all the flooding, so they decided to turn around and return to their motel. As they traveled back to their motel, the radio alerted them to local tornado warnings. They eventually made it back to the location of the motel, exhausted and disillusioned, only to find that the motel had been hit by a tornado and was no longer standing.

Mike's wife decided she had enough of Florida's weather and told Mike she didn't think they could move to Florida. They drove as fast as they could to the East Coast and stayed for two days to recuperate from the trauma of the tornado before heading home to Atlanta.

Bobby understood their decision to stay away from Florida considering their experience, and it would be another six years before Mike would join the Arch team. When he finally did, it was on safer ground in Atlanta.[1]

take up shop from that point on would be determined by location, location, location!

Leon turned his attention to growing Arch's aluminum business. In the early 1990s, Steve Kendall, owner of the nationally known, New Jersey–based aluminum manufacturer Aluminum Shapes, had befriended Bobby and his partner, Marvin Cravetz, and begun selling them on the idea of a joint venture. Arch needed capital to fund its new vision of an ever-expanding Arch, and it knew Aluminum Shapes would vouch for Arch when it needed access to loans, so the Silversteins struck up a relationship with the company.

During the initial courtship, Arch had borrowed money from a bank that had an alliance with Aluminum Shapes. The bank had financed Arch's capital expenditures and expansions, such as the Sunshine Laminating buyout, because of Arch's connection with Aluminum Shapes. Now, to further the aluminum business, Arch resumed its discussion with Aluminum Shapes over a potential merger.

Also, at the time, Amarlite, the company that Sydney and Bobby had dealt with in the 1960s, had gone into bankruptcy. Leon saw the purchase of Amarlite as a great opportunity to launch Arch to even greater heights of national exposure. "We were underfunded and also considered a schlock company. There were always rumors about our inability to pay our bills. In fact, to temper these rumors, we started our own rumor that Marvin Cravetz, one of the original partners, had won the lottery," Leon recalled.[35]

Playing an artful game with the bankruptcy process, Leon approached Congress Bank about purchasing Amarlite and placed a bid. Once the bid had been entered, the business would be sold at auction, as required by law. Leon and Bobby placed special conditions on the deal so that anyone bidding to buy Amarlite would not be prepared to meet the terms at the auction. Another group did outbid Arch, but when the judge asked if they had funds available for immediate closing, the answer was "No." With that, the judge hammered on the desk and declared Arch the new owners of Amarlite.

It was, once again, the relationship with Aluminum Shapes that made the purchase possible. Aluminum Shapes' wealth and relationship with banks helped Arch fund the project, which brought the entire Amarlite product line over to Arch. Leon explained how this seemingly straightforward acquisition quickly grew more complex:

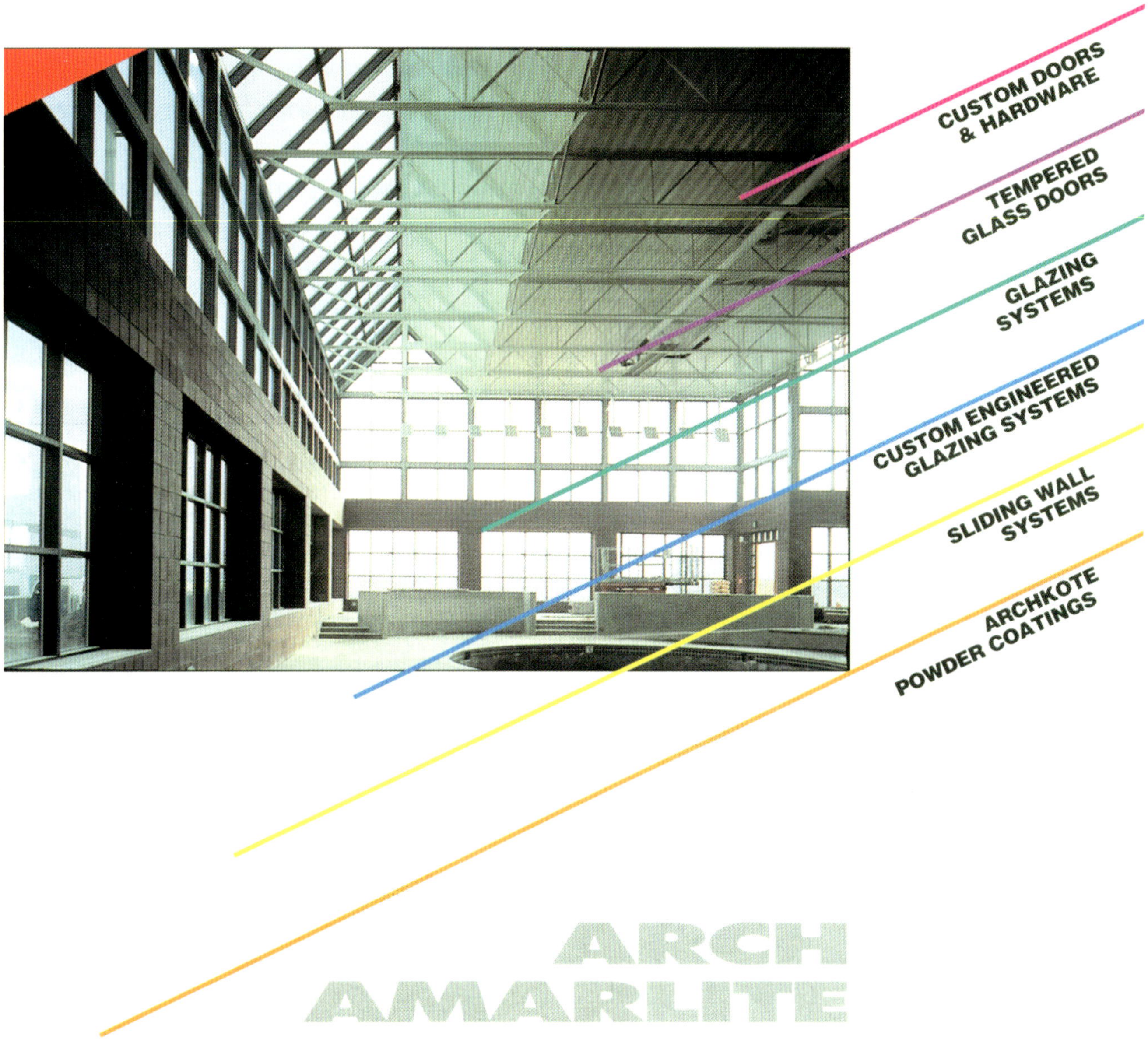

After acquiring Amarlite, Arch created this brochure to publicize the Arch name and to promote the new product line available from Amarlite.

Amarlite had facilities all over the country. We ended up reselling the Dallas facility and its assets to our cousin Ray London in Dallas. The Cleveland facility we resold to the Perilsteins in Pittsburgh. But with the assets of Amarlite, we also bought all the junk with it. We had to figure out how to get rid of as much as we could because all we really wanted was the product line [and brand]. Amarlite was in complete disarray, and the business needed to consolidate from seven national facilities to just two.[36]

An enormous amount of material, product, and equipment also required transportation across the country. Because of overdue bills, landlords for some of Amarlite facilities threatened to padlock their warehouses. However, despite all the problems initially associated with the Amarlite deal, the purchase turned into a great boon for Arch. Leon elaborated:

> *We bought the product line, and they were probably doing $20 to $25 million at that time. That was a pretty big deal for us, because as big as Arch was starting to get, every architect in the country knew Amarlite. Amarlite was in every Wal-Mart spec, Target spec, and Kmart spec. If you went to an architect and said, "You know, we're approved by Wal-Mart or Kmart and Amarlite," it was fine. If you said, "Arch Aluminum," they said, "Arch who?"*
>
> *Except for the place in Indianapolis, we were really a Florida-based company, and buying Amarlite put us on the map. We became a household [name], and that was really the point. We became much more visible in the metal field because of that acquisition. [Initially we] lost a lot of money ... but that was one of the few acquisitions where the payoff was three, and four, and five years down the road.*[37]

Following the purchase of the Amarlite assets, Arch heard that a small glass fabricator called Hardtemp, a division of Harding, located in a good neighborhood in Kansas City, was up for sale. Leon flew out to speak with the owners. During the meeting, Leon asked a lot of questions, and seriously considered making an offer.

But first he left the room for three minutes to call Bobby. "Dad," Leon said, "We just bought Amarlite. We are just opening up, and we've got all these things going on. We have no money, but I'm going to go in there and tell them, 'Let's make a deal.' But I need to be able to buy the furnace because they don't own it."[38] They were selling the company for depreciated assets, making it less expensive, but part of the deal needed to include the furnace and the lease. "The bank owns it," Leon said, "and I don't want a place without a furnace. Their lease is up in six months. I'm going to tell them I need to talk to the landlord and renegotiate a new lease for the next five years."[39] It took Leon another five minutes of meeting with the owners to seal the deal with a handshake. Arch took over the plant, purchased the oven, and got a new lease 30 days later.

This 1995 purchase was a milestone for Arch. It was the first time that Leon and Bobby purchased a plant that was already up and running. Arch obtained the property fully staffed and operational. To ensure that there was no downtime, Leon and Bobby spoke with the employees prior to closing on the deal. Leon explained to them how they planned to close down as Harding and reopen as Arch. Everyone would be let go on Friday and rehired on Monday. As promised, the plant closed under the old management on a Friday, and the employees lost their jobs. And on Monday morning, Arch opened its doors and hired everyone back.

"It was a good experience. We lost some money the first year and had some issues, but, overall, it was a [good] formula," Leon said. "We were able to get into a new region, a new location, with minimal capital outlay, and then over the next two years, [we invested] a lot of money in new equipment and brought the place up-to-date."[40]

A Promising Merger

Based upon their positive relationship nurtured over the past four or five years, Bobby and Leon decided to merge with Aluminum Shapes. Stephen Kendall, president of Aluminum Shapes, would become chairman of the merged company, Bobby would become vice chairman, and Leon would serve as a member of the board.

Since 1954, Aluminum Shapes had claimed the position as the largest privately held aluminum extruder in the country. At the same time, Arch had become the largest architectural aluminum and glass fabricator in the country. As a united entity, Arch and Aluminum Shapes would approach sales of $300 million. Based upon the sales projections, a merger of these two companies looked profitable for both businesses. But, the merger would not be completed until 1996, when what seemed like a bright future suddenly turned dark and ominous.

Arch literature from the late 1990s. These catalogs indicate Arch's expanded product line and reinforce its hallmark single-source shop concept.

CHAPTER SIX

ON THE EDGE OF THE PRECIPICE

1996–1999

What Bobby and Leon built was a good, solid platform, and I think Arch was poised for more growth.

—Richard Jaffe

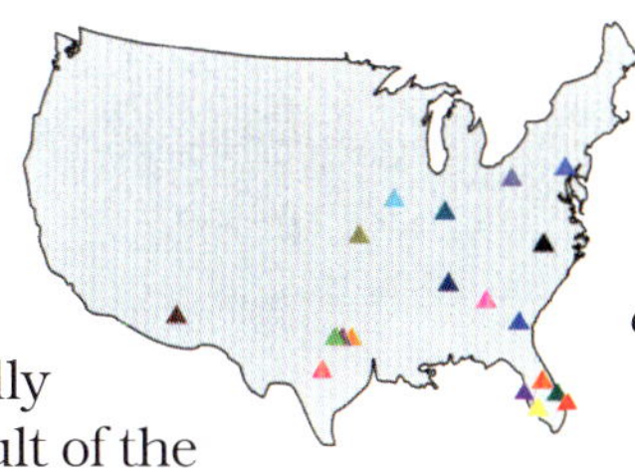

THE YEAR 1996 BEGAN WITH everyone's hopes soaring. First, Bobby and Leon implemented their plan to position Arch strategically in the national arena. Second, as a result of the impending merger with Aluminum Shapes, Arch could pay down its enormous debt. And third, Arch Aluminum & Glass became the first company to earn approval by Miami-Dade County for hurricane product certification. That certification gave Arch a good four-year head start on fabricating impact-resistant commercial glass and aluminum products.

On January 1, 1996, Arch Aluminum & Glass and Aluminum Shapes, along with two other entities owned by Stephen Kendall, formed a new business entity called Arch America. Under this agreement, each company would remain completely autonomous and manage its operations independently.[1]

Arch had always been profitable, but because Leon and Bobby reinvested everything back into the company, it had remained cash poor. The merger changed that, providing Arch with access to financial resources, which allowed it to eliminate the debt it had accumulated over the last several years. Unfortunately, the merger, soundly worked out on paper, would suddenly turn sour.

Within a week of signing the papers, the problems began. Leon explained:

Kendall acted like it was a takeover. He reminded us constantly that he was the majority shareholder and that he was the boss, and therefore he could dictate policy and procedures to us. But we were not going to do anything that we did not think was in the best interest of, or the right thing to do for, our company.

Kendall demanded that we change the way we managed our business. He insisted that his company, Aluminum Shapes, exclusively provide the aluminum for Arch Aluminum & Glass.[2]

Buying exclusively from Kendall remained a key point of contention for the Silversteins. "Aluminum Shapes had long lead times, and, in some cases, supplied inferior products to those [of Arch's other suppliers]," Leon explained. "Arch always sold itself on service, but how can you provide service if you can't get the [extrusions] and what you do get is of poor quality?"[3]

Also, Kendall's management style ran counter to Arch's hallmark down-to-earth style. According to Leon, Kendall ran his business from the top down. He made the decisions, he gave the orders, and everyone else did his bidding. "So everything he wanted us to do, we did the exact opposite," Leon recalled.[4]

This map illustrates Arch's growing presence across the country as of 1997. Arch would extend its reach even farther in the coming years.

Within the first month of the merger, and in a moment of unbearable frustration, Leon drew up his resignation. "Then I took a moment to consider how my resignation would affect my dad, my brother, Kevin [Aubuchon], and the rest of my family," he explained.[5] Leon would remain on, determined to fight.

Leon often became argumentative because of his frustration: "If the wall was white, Kendall would say the wall was blue. He expected everyone to say the wall was blue. I would say, 'You are wrong. The wall is white.' He couldn't take criticism. Kendall told my parents I had psychological problems."[6]

The Pressure Mounts

Lawsuits resulted as the animosity between the partners grew more tangible. Leon explained: "Kendall would fly down to Florida from New Jersey and ask us, 'Are you in or are you out?' I would respond, 'I am working, what more do you want?'"[7] To this day, the Silversteins joke about the question. But it also serves as an unpleasant reminder of those perilous times.

Over the years, Kendall maintained relationships with the teamsters union, and he wanted to bring labor union officials to one of the Arch operations in New Jersey. "At the time, we asked the employees to vote for unionization, as a favor to Kendall. So the employees did," Leon explained. "But they soon found that unionization generally meant the addition of rules, structured salaries, less freedom for upward mobility, and indeed less benefits."[8] The employees would reverse their decision to unionize in the near future.

According to Leon, Kendall tried another approach to get Arch under his control: "He sent a few rough-looking men down to Florida to remove us from our facility. I believe Kendall hired these men to physically intimidate us and scare us into handing the business over to him. Unfortunately for Kendall, the people he sent never made it into the facility because we got wind of the plan. We locked the entranceway and called the police."[9]

A New Tactic

"Kendall continued to argue that we did not operate Arch in the best interests of the larger entity, Arch America," Leon explained. "Kendall believed that this contention was grounds for a hostile

INTO THE FIRE

DON HYDE, WHO WAS HIRED IN 1996 to oversee finances at Arch, arrived at a crucial and difficult time. "There were a lot of firsts when I arrived. We were having our first audit, our first budget, our first finance meeting."

At the time, the company needed to raise $60 million in financing to divest itself from Arch America. Hyde said, "It was a harrowing but fun time. Everyone was pushed to the limits of their abilities and beyond." To secure the funds, Hyde focused on working with public financial institutions, while Leon networked with private investors. Hyde explained:

> *On a Friday, the underwriter from one bank told me it was a done deal. He said that his bank could supply Arch with the necessary funding. On Monday morning, he called me back and said, "I'm really sorry, no deal." I'm sure there was someone inside sabotaging the deal. But we never could prove it.*[1]

removal of the Silverstein family. We disputed his argument and maintained that the original agreement guaranteed autonomy to all parties, and that Kendall violated the deal."[10]

While the discord with Kendall continued, the Silversteins recognized the need to make management changes to support Arch's newfound national presence. However, they also strove to maintain the mom-and-pop atmosphere they cherished. Leon fine-tuned operations and implemented new management strategies at the facilities to make operations consistent from one branch to another. He also searched for top management who bought into the Arch culture and its operational approach.

In the meantime, despite the fighting with Aluminum Shapes, Arch thrived. Because the two companies were still connected and the parent company was financially strong, Arch didn't have to worry about money pressures for the moment. This support bought Arch time, allowing it to use these resources to become more profitable. "That merger was the worst thing and the best thing that happened to us," Leon noted.[11]

In 1996, due to its expanding business, Arch outgrew its headquarters in Pompano. The Silversteins decided to move the business to an area that offered business incentives to move. Also, because Leon, Bobby, Ricky, and Kevin all lived in different cities,

Opposite: Construction begins at the new Arch facility in Tamarac by breaking ground, then digging a lake with a series of canals to accommodate the water runoff resulting from building in the wetlands.

Above left: The Tamarac facility was built by using prefabricated building pieces to provide for rapid construction and to avoid weather-related delays.

Above right: With the wall and roofing in place, interior construction begins.

Right: Nearing the end of construction, the building quickly takes shape with the final installation of windows.

they decided to look for an area that would be centrally located for everyone. They found the best of both worlds in Tamarac, Florida.

At the time, the city of Tamarac was eager to create an industrial park. The city commissioners wooed Arch to come to their city because they believed the company's presence in its new industrial park would entice other businesses to move in as well. As a result, the city gave Arch $100,000 in relocation money and waived the permit fees. While the Silversteins did not believe the incentives provided Arch with a significant amount of money, they still found the move to be strategic, based on the location. The new facility would be centrally located and have access to major highways. The extra money was just an added bonus. "It is one of the best things we have ever done," Leon commented.[12] Arch became the first commercial building in the Tamarac industrial park. Several years later, Arch constructed a second building across the street from the original plant.

Branching Out

The facility in Austell, Georgia, successfully provided the busy Atlanta market with service, but Leon wanted to add a tempering oven. Because the plant was too small to accommodate one, Arch planned to move the facility. The new facility needed a strategic location to properly serve the Atlanta market. Using the lessons it learned from its experiences in Indianapolis and Iowa, Arch contacted the economic development people in the areas surrounding Atlanta for assistance. With their input, Arch ultimately decided Villa Rica would best satisfy its needs and the needs of the Atlanta market.

"We were going farther and farther away from Atlanta. We were going about 20 miles west of Austell, and we were afraid we would lose some of our customers on the east side of Atlanta," Leon explained.[13] So, Leon and Bobby decided to open a satellite distribution facility in Norcross, Georgia, to be a depot for the smaller-sized customers in the area. "We rented a 10,000-square-foot facility on the eastern side of Atlanta so we would be able to deliver to our east," Leon said.[14]

Customers on the east side of Atlanta would now be able to pick up materials on the east side without having to drive across town to Villa Rica.

Arch completed the move to Villa Rica in 1996 and, less than a year later, opened a new plant in Youngsville, North Carolina. Youngsville, like the one in Iowa, was a greenfield plant—a plant started

Above: This 18-wheeler, parked at the Villa Rica facility in 1996, is a far cry from the original small truck Bobby used when he opened the doors to his fledgling business in Miami.

Left: Leon decided to build the new Villa Rica facility to accommodate one of the new tempering ovens. Through careful planning, this facility successfully serves the needs of the busy Atlanta market.

from the ground up. Leon commented, "In Iowa, we learned how to talk to the economic development people. We contacted the economic development people in different municipalities around Raleigh and ended up settling in Youngsville. It was a growing and booming area. We immediately started doing very well in Villa Rica. We'd been in Atlanta a long time, and we were trucking product farther and farther north into Charlotte. So, Youngsville was just a natural extension of Villa Rica."[15]

Above: The family grew Arch aggressively in the mid-1990s. In less than two years, Arch built the Villa Rica plant, opened the distribution center in Norcross, and built this facility in Youngsville.

Below: (From left to right) Claire Murphy, Kevin Aubuchon, and his wife, Donna, were instrumental members of the IT team who developed custom software for Arch and trained employees across the country on how to utilize the new system effectively. *(Photo courtesy of IDS Photography.)*

Technology Takes Hold

As Arch continued to expand, each new facility required computerization to optimize production and to network with Arch's numerous plants. Kevin Aubuchon oversaw the computer implementation at all the facilities including Villa Rica, Youngsville, and Norcross.

Before being transferred from Orlando to Atlanta to assist Aubuchon with computer technology, Don Powers performed a variety of duties at Arch. Powers recalled how he moved from working in the warehouse to assisting with information technology (IT):

> *The first week I worked at Arch, I worked 79 hours. I started just breaking out glass. From there, any department I wanted to get into, I did. I ran a cutting table. I fabricated. I moved into the office and started working on production reporting. We set up reporting so that we could figure out what was going on in the plant. I was pretty good with computers. They moved me inside, and once they moved me inside, I really understood. I learned the whole process of how our facilities run, and how they should be managed.*
>
> *As Leon expanded the company, I was sent out to other facilities. It was easy to go out there and be a hands-on type of guy, to show somebody how to run a cutting table or unload a truck. I was traveling and setting up facilities. For about two years straight, I traveled for them.*[16]

Planning for Success

Each Arch acquisition follows a general pattern and establishes a similar culture. "There's a feel," said Leon. "You can tell it's an Arch place." After Arch purchases or secures the assets from an existing facility, Arch devises an action plan with specific goals to guide the facility to increased sales and profitability. Arch assigns a new management team to the facility and closely monitors its performance in the areas of production, sales, and profitability. The facility receives any necessary equipment upgrades and increased sales support it requires to provide the appropriate infrastructure to ensure success.

If the facility is unable to meet the outlined goals within a predetermined time frame, Leon and the management team may decide to close the facility or launch an aggressive improvement plan with specific target dates to address the underperformance.

To date, only two facilities have been closed. In each case, Arch successfully opened a new facility to meet the area's market needs. Arch opened a site in Trident, California, to replace the Portland, Oregon, site, and after closing the Memphis, Tennessee, site, Arch opened one in Jackson, Mississippi.[1]

Powers eventually went to Indianapolis and became plant manager there. Years later, he would move again to an Arch facility in Akron, Ohio.

Another member of the information technology team, Claire Murphy, who had joined Arch in 1994, traveled to all Arch branches to make sure the plant managers and other employees used the computer systems effectively. "Once all the programs were in place, we were able to start doing things for the whole company. It was the first step in consolidating all the information," Murphy recalled. "Up until that time, orders were keyed into a branch facility, and then hard copies were made and sent by mail down to headquarters in Tamarac. There was no interconnectivity."[17]

Murphy investigated the purchase of a Hewlett-Packard computer server to help facilitate the necessary information technology changes and presented her proposal to the family. Another employee went over Murphy's head and, on his own accord, brought in a quote for the same item for $10,000 less. However, Leon and Kevin signed off on Murphy's proposal, despite the additional cost. "I felt supported, and [realized the extent of] their loyalty," Murphy said. "It was then that I knew I was really at home at Arch."[18]

In addition to hiring Murphy, Aubuchon brought in Terri Weber to join his team as a computer programmer. According to Murphy:

> *Terri made us look at how we were operating our IT systems. There are two ways to behave in business; one of them is to fly by the seat of your pants, solving problems of the moment and not planning, and the other way is to take a long look at the picture and decide where you are going and to be conservative about it. Programming is the same way. You can go problem by problem, or you can ask, "Where do I want to be 10 years from now?" In actuality, if you fly too much by the seat of your pants, it costs you money because you haven't planned.*[19]

Having purchased a server and other equipment earlier, the information technology department began to coordinate an advanced internal computer system with an Intranet (a network accessible only to Arch employees). The initial process involved connecting all the plants over phone lines. In addition, the information technology team built in a good deal of safety checks to ensure that potential system failures in one place would not interfere with the overall business operations.

While upgrading systems in existing and new facilities, the IT team also trained employees and helped troubleshoot computer issues. "We wanted to teach people how we expected them to do business at Arch," Murphy explained.[20] To unfold the new technological advancements effectively, the team implemented processes bit by bit. "The IT management team is pragmatic. We were not going to overdo the computer technology. We were not going to do fancy things," Murphy said. "We did the transition in a smart fashion. We did it slowly, until we got everyone on board."[21]

Reflecting upon the family's response to the computerization, Kevin explained:

> *Like father, like son, like brother. The computer systems are not something they really want to learn. It is part of their business, so they use the system. We don't bother training them in the finer details, the specifics of the environment. They don't want to look at a computer all day. That is a good thing. It's a nice mix between the young manager who has grown up in the environment where he is used to sitting down and staring at the computer to see how his business is doing versus the guy who is going to walk out on the shop floor to see how his business is doing. There is a big difference.*[22]

Free at Last

Despite Arch's success at quickly expanding nationally and making great strides at building the brand name, a dark cloud still hung over Arch; the Aluminum Shapes dispute lingered on.

PRODUCTIVITY ON THE RISE

THE CHART BELOW DEPICTS THE STEADY growth that occurred within the glass industry from the late 1980s until the turn of the century. Labor productivity statistics rose in every area related to glass production from 1987 to 1999. Though productivity increased, the industry actually utilizes fewer workers now than in the past due to computer technology advancements.[1]

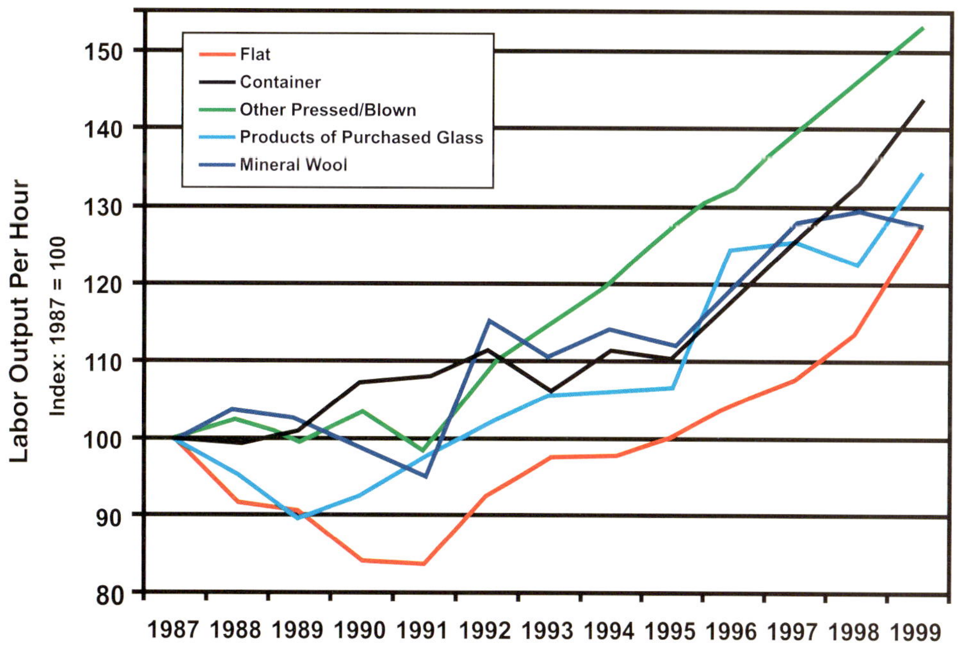

Chart courtesy of "Glass Industry Analysis Briefs," Energy Information Administration, Office of Industrial Technologies, U.S. Department of Energy.

Finally, in 1998, Arch and Aluminum Shapes came to a settlement to end the animosity and costly fighting. Leon explained:

> *Kendall agreed to give us 120 days to raise the money to pay for all the debt that we had on Arch's books, including a closing fee for Aluminum Shapes. If we were unable to raise the money in 120 days, then we would be forced to sell Arch at a ridiculously reduced price, and [the Silversteins] would be forced out of business [with a five-year non-compete clause]. We really had a gun to our heads.*[23]

MEMORY LANE

BOBBY AND BEANI'S YOUNGEST daughter, Cindy, recalled Bobby's love of cars and his propensity to make jokes:

> *It was always a Mercedes, after Mercedes, after Mercedes. He would go outside and shine his car after work. He took pride in that car, major pride in that car.*
>
> *Every time I speak with my oldest friend, she brings up the story about my dad, his car, and the rubber hands. He'd hang these rubber hands from the trunk of his car. My dad thought it was very funny. But you know, he was always a very "quiet" funny. He didn't laugh and smile. ... He would just look straight-faced and wait for the response.*[1]

The Silversteins agreed to the stipulations and set out to find individual investors or other means to finance the purchase. "We had two banks involved because we couldn't afford to have one bank drop out at the 11th hour," Leon recalled.[24]

Despite all the planning and precautions they took, the Silversteins found themselves several million dollars short as their deadline rapidly approached. Finally, Marvin Cravetz, one of Arch's original investors, succeeded in finding an individual investor, Steven Marcus, who put up the necessary cash just in the nick of time.

According to Leon, Marcus said, "You are good people. I like to do business with good people."[25] So with Marcus' backing, Leon and Bobby felt vindicated because they actually pulled off the deal and got the necessary funding in time.

A deal is a deal, and Arch was now free from its counterproductive relationship with Aluminum Shapes. Having extricated themselves from their troublesome and frustrating position, Leon and Bobby could focus their energies on running a successful business. Over the next nine months, Arch again experienced tremendous financial growth.

Meanwhile, the mayor of Jackson, Mississippi, approached the Silversteins with the idea of bringing Arch Aluminum & Glass to his city. Despite problems with the Memphis facility, Leon still wanted a presence in that part of the country. "We never made money in Memphis, but we still had a customer base," Leon explained. "The market was starting to come back, and we were trucking from Villa Rica, Georgia. Jackson was a nice, quiet city. Harold Jackson, the mayor, offered us a deal on land. He convinced us and helped us understand that Jackson was a place we needed to be."[26]

Opposite left: All of Arch's tempering ovens are the flat-horizontal type. The assembly of the ovens takes place on-site because of their large size. Here, the base of the oven has been constructed.

Opposite right: A worker assembles various oven components.

Above left: The tempering oven's heating and cooling process is critical to the successful creation of glass. This picture depicts the blowers and ductwork that help control the temperature of the glass.

Above right: The fully installed oven as it appears at the Jackson, Mississippi, facility.

The brokering deal in Jackson ran smoothly due to the Silversteins' astute business acumen gained through negotiations for the plants in Indianapolis, Iowa, Tamarac, and Villa Rica. They had come to understand how to work with city officials to benefit all sides.

With Arch's Memphis lease about to expire, the labor union brought a suit to its labor relation board, forcing Arch to unionize that facility. However, the family was not concerned because they knew that the plant would close shortly anyway. Arch had decided nine to 10 months in advance to open in Jackson but had kept the opening quiet, hoping to keep continuity. At the same time, the new facility was being built with new equipment and new people. When the lease was up, Arch closed the Memphis plant and moved to Jackson. But the union again tried to unionize the new plant. The union's efforts failed, though, due to a stipulation that prevents unions from following a company to a new location. As a result, the Jackson plant remained nonunion.

As a greenfield project, the Jackson facility was built from the ground up with all new equipment, including a new tempering oven. "When we moved from Memphis, we had a new place all set up to start running fully operational," Leon remembered. "We turned off the lights on Friday in Memphis, and we opened in Jackson on Monday. Our customers had tempering, insulating, and service without interruption."[27]

Seeking the most economical way to produce product and to move it through the warehouse, Leon always found the best possible use for existing equipment. He also sought to keep all plants fully operational despite moves. "We always plan two moves at the same time [so there is no down time]. In 1996, we moved into Villa Rica and we also moved into Tamarac, Florida. The equipment we had in Villa Rica came down to Tamarac, and the equipment from Pompano was sold to Beani's cousin, Ray London, in Texas," Leon recalled.[28]

A Time of Reckoning

Steve Marcus, the man who provided the Silversteins with the much-needed cash to buy back the business from Aluminum Shapes, decided to convert his share of the Arch investment into stock. Now a major shareholder, Marcus, along with Bobby's original partner Marvin Cravetz, who still owned a good portion of the company, hoped to convince Bobby to sell Arch. Bobby started to worry he would lose the company he worked so hard to build, since he had continuously reinvested resources back into the business.

Leon, however, was confident they could secure the necessary funding to prevent Marcus and Cravetz from forcing an outright sale of the company. Besides having an intuitive understanding of the market, Leon possessed shrewd negotiating skills, as well as unparalleled drive. The high stakes served to motivate the family further as they prepared to battle Cravetz and Marcus. When they finally arrived for the showdown, Leon was ready for the fight.

Marcus and Cravetz had hired Richard Jaffe, an attorney from Philadelphia, to represent them. Jaffe explained:

> *Marcus, an existing client of mine, called me one day and indicated that he was a shareholder in Arch Aluminum. He wanted to sell his investment. He also talked about selling the company. He asked me to represent him. I didn't know the Silversteins at that point.*[29]

The trio met with Bobby and Leon to decide the future of Arch. Jaffe was very impressed with the Silversteins, and during their initial conversation, Jaffe made a positive impression upon Leon. That meeting, however, was far from satisfying for all the parties involved.

In an effort to thwart Marcus and Cravetz's efforts, Leon made use of bluffing skills. At the meeting, Leon said: "You want to sell, fine. But you are not selling with me. I don't think you can sell the company without the management."[30] It was obvious the business would not sell for as much money without the management included, and the discussion heated up.

Then, unexpectedly, Leon and Bobby turned the tables on the two of them. Leon explained:

> *I said, "You want to sell. I want to buy." Then Marcus quipped, "You can't buy it. You don't have any money." I said, "No one is going to buy this business knowing that I will just take the money from the sale and start up a new business all over again, and no one is going to make us go back into business across the street." That's exactly what we did earlier in the 1990s, as a result of our dispute with Westshore Glass.*[31]

The atmosphere in the conference room grew tense, with Leon and Bobby seated on one side of the table and Marcus, Cravetz, and Jaffe on the other. In response to Leon's measured threat, Marcus demanded that Jaffe tell Leon he had to sell and stay on. However, Jaffe had found Leon's argument to be beneficial to both sides, which angered his clients. Finally, Jaffe stood up and switched sides. Jaffe recalled how Leon responded:

> *I remember Leon actually said to my client, "I would like to have Jaffe represent the family and have you get another lawyer." I helped them structure the purchase of the business from those two shareholders. Marcus and Cravetz found other counsel.*[32]

To secure the funding necessary to finalize the deal, Leon and Jaffe went to several lenders. Leon expected to involve more than one bank, and he

ONE JOB

ORPHA LOPEZ STARTED WORKING for Arch Aluminum & Glass in 1988, while she attended night school in Miami. The commute was long and tiring, but Orpha enjoyed her job and liked working with the Silversteins and their customers. She started out in sales, then moved to purchasing, and did a little bit of everything in between, until she started working for Leon in exports 12 years ago.

Orpha explained why she stayed at Arch all these years:

> *We treat our customers the right way. This is my job, and I have to take care of my job. I have always done what is best for Arch. I love coming to work. It's been a long time, but I feel comfortable. I go to work in peace. I like to be here. I'm sure it's the one and only job I will ever have.*[1]

Opposite: At the 1997 National Glass Association show, Arch displayed its large array of products through this giant display that would be utilized successfully over the next several years.

ALUMINUM & GLASS L.C.
MANUFACTURERS OF:
AMARLITE ARCHITECTURAL PRODUCTS

WHAT YOU NEED TO KNOW

GERRY BOYLAN, ONE OF THE OWNERS of Long Point Capital, explained why he wanted his company to invest in Arch: "The quality of the people determined the benefit of the relationship and the profitability of the relationship. Our assessment of the Silverstein family was that we were dealing with ethical and honest people. We thought that minimized the risk significantly."

In regard to his relationship with Bobby, Boylan recalled:

Bobby would sit behind his desk and say, "Gerry, let me tell you what you need to know." I enjoyed these talks and made a point of sitting down to speak alone with Bobby on every visit. At every one of those meetings, I learned something. It wasn't necessarily about Arch Aluminum & Glass, but it was useful information about business. I found Bobby to be a great mentor for a family business. He was very thoughtful about how to work with his family and employees.[1]

sought investors willing to grow the company, take risks, and serve as silent partners—allowing the Silversteins to run the business as their own. A tall order to fill, but one fully satisfied by Long Point Capital, a private equity company.

Recommendations Pay Off

One day in 1998, Gerry Boylan, principal of Long Point Capital, received a telephone call from an investor colleague, Lother Mayer, a friend of Bobby's. Boylan explained:

Mayer called me and described Bobby and Arch Aluminum & Glass as a wonderful company run by wonderful people. He suggested that Long Point consider investing in Arch. That same week, Gale Lemer, owner of Gale Industries and another Long Point investor who knew Bobby, put a call through to me. He said, "Gerry, I have to tell you about Arch Aluminum & Glass, a company that you should buy.

That is a rare event in our world, to get two such recommendations in the same week. So, I called Arch and spoke with Leon. The conversation was refreshing. Leon explained that Arch Aluminum & Glass was sold to another company and that the partnership was not good, and they wanted to get back to running a family business. He also told me that he was going to pick the partner they were going to do business with, as opposed to the traditional way, in which we go in and pick who we want to work with. I understood right away that I was dealing with very strong-minded entrepreneurs. I appreciated Leon's directness. That also seems to be a family trait.[33]

Impressed with Leon and his forthright manner, Boylan and his partner Ira Star flew down to Tamarac to meet the Silversteins and gain insight into Arch. "The family management team placed a high value on loyalty, hard work, being competitive, and winning," Boylan said. "All family businesses have unique dynamics—sometimes positive, sometimes negative—but we found that the Silverstein family was a highly functional family business, and they knew how to work together. ... The family makes unusual efforts to work with the employees who they consider to be extended family."[34]

The meeting resulted in a business partnership between Arch and Long Point Capital. On September 30, 1998, with Long Point's financial assistance, Leon and Bobby formally bought out their former financial partners. Cravetz, who had invested $75,000 and loaned Bobby another $75,000 back in 1978, received $8 million in the buyout.

In summing up the arrangement with Arch, Jaffe said, "The deal with Long Point Capital, including placement of two of Long Point's investors on the board of directors at Arch, further streamlined the approach to running a substantial-size business."[35] Long Point brought more sophistication and discipline to the operation of the business,

Life's Work

Bridgio Lopez came to work for Bobby in 1978, just a few months after Arch Aluminum & Glass opened its doors. Prior to working at Arch, Bridgio delivered glass for another company, but in his travels, he had heard complimentary stories about Bobby and Arch.

One day, his job brought him to Arch to pick up materials. While there, he spoke to Arnie Studenski about job possibilities. Bridgio told him, "I can build anything you need me to build." Arnie took him to see Natalie Reed, Bobby's sister, who manages the Human Resources Department for Arch. She gave Bridgio an application and hired him that same day. It was a Friday, and she told him to come to work on Monday. Bridgio said, "No, I want to start today. I want to start now."

Reed agreed, and Bridgio started that day, successfully building Studenski a toolbox with a lock for the warehouse by the end of the day. Because of his skills as a carpenter, Bridgio receives the more complicated design projects. Sometimes he catches errors in the architect's plans and makes alterations to fix them. Bridgio said, "I want our customers satisfied with everything I make. I have no plans [to retire]; I plan on working until my last day."[1]

and the Silversteins learned a more formal approach to business.

Reflecting on the late 1990s, Leon noted: "We needed to be better managed. The arrangement also solved the problem of having two dissident shareholders who were not interested in growing the business."[36]

Limitless Potential

Despite the stress associated with the latest hostile events, Arch, with its ever-expanding facilities, flourished. During the 1990s, Arch successfully opened two mirror plants, two distribution centers, and two new facilities. Additionally, Arch relocated to larger facilities in four locations. It also acquired two glass and aluminum plants. Only the Memphis, Tennessee, plant proved an unsuccessful venture.

By the end of the 1990s, Leon had sharpened his business skills in all areas, including initiating acquisitions, managing relocations, buying equipment, and negotiating complicated deals. As Leon grew comfortable and more confident in his role, he started accepting assistance from others and improved his interpersonal skills. As a result, Arch benefited from the expertise of others who joined the management team, and Leon was able to focus on formulating a strategic plan and a strong vision for the company. Ricky, the younger Silverstein, who, up to this point, spent his time with the company in sales, soon would take the helm as chief of operations.

As the 20th century came to a close, Kevin Aubuchon searched for a computer program that would effectively address issues that could arise in regard to the Y2K bug. After looking at various existing software packages, Kevin decided that the IT staff would write its own software package to meet the exacting requirements for Arch's operations. "Having the IT team create its own software was a pretty big milestone for Arch," Leon said.[37]

Through it all, Arch successfully transformed itself from a small mom-and-pop operation to a large national company. With many business operations in place, recruiting new talent was key to maintaining Arch's ever-evolving leadership role in the glass and aluminum industry.

Moving forward, Arch would face new obstacles, and the family would experience problems inherent with growth. Industry issues—such as the need to develop energy-efficient processes and green processes, and meet industry standards—would further challenge Arch and its dedicated team of employees. But providing superior service and creating quality products remained paramount.

This creative 2005 advertisement plays off the *Men in Black* movie theme and markets Arch and its branch managers to the architectural glass industry.

CHAPTER SEVEN

INNOVATION AND EXPANSION

2000–PRESENT

There is now a method to our madness.

—Leon Silverstein

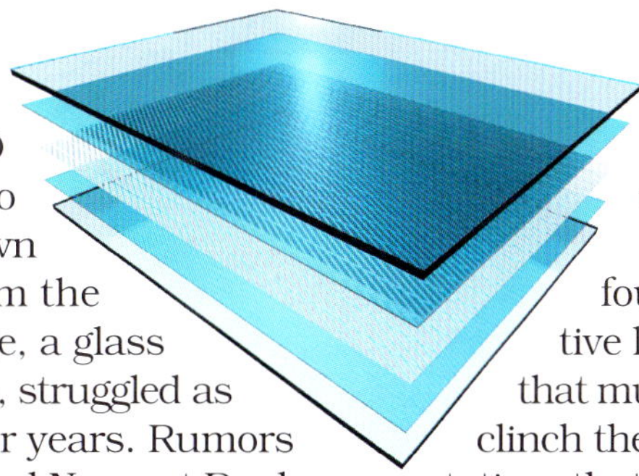

THE YEAR 2000 PRESENTED opportunities for Arch to expand despite the slowdown in the economy that resulted from the bursting dot-com bubble. Coolidge, a glass fabricator in Waukesha, Wisconsin, struggled as an automotive glass distributor for years. Rumors spread that the plant had closed and Norwest Bank had foreclosed on the assets. Official word of such happenings takes time to surface, but Arch learned about the closing from some of its glass supplier contacts. Leon believed that buying the Waukesha plant would be a perfect opportunity for Arch to expand farther into the Midwest market. He quickly contacted the bank and indicated the company's interest in buying Coolidge's assets.

Although two other potential buyers existed, Leon once again found himself in a position of strength. As CEO of Arch, he could make the deal then and there, if he chose to do so. "All the bank wanted was to know how much and how fast. They were only concerned with the loan and not the other creditors," Leon explained. "I literally sat there for three days, knowing that while I was there, the bank wouldn't allow someone else to look at the assets. I sat there, and I tried to figure out what kind of a deal I could make."[1]

The bank wanted to sell the assets of both the Coolidge flat glass operations and its automotive product line. While Leon planned to purchase the flat glass and insulated glass equipment, he did not want the automotive products.

Fortunately for Arch, the bank found a buyer for Coolidge's automotive line, moving the Coolidge purchase that much closer to reality for Arch. To help clinch the deal, Leon told the bank representatives that he would assist them with collecting the receivables. The bank agreed to Arch's terms. "After the purchase, we resolved issues with the previous Coolidge customers who didn't want to pay their bills. If a customer complained about a piece of Coolidge glass, we would replace it. We also honored any warranty work," Leon noted. "I knew we could make up the cost once the customers started reordering from us. Finally, we bought the receivables that were left."[2]

Most of Coolidge's assets were not owned by the company, but rather leased, including most of the manufacturing equipment and office furniture. Leon became deluged with angry calls and demands for payment from leasing companies. "We negotiated one by one. If we could get a fair price, we would buy the lease out," he explained. "In about 30 percent of the cases, we couldn't agree on a

This artist's rendering depicts the many layers of glass and materials required to create a single piece of laminated glass.

From left: Glass division employees Arnoldo Giallego, Lauto Arce, Horacio Cruz, and Jahir Valencia show off some of the special protective gear worn as a safety measure when fabricating glass. *(Photo courtesy of IDS Photography.)*

value. In those cases, we told the leasing company to come and take their equipment."[3]

Leon cited one example in which he offered to buy $150,000 worth of furniture from the leasing company, but the company refused his offer. "I said, 'In that case, come get your furniture, and remember to bring a truck and people. It's not my furniture, and I'm not loading your truck,'" Leon said.[4]

In other cases, Leon negotiated deals that were very favorable to Arch, one of which included a robotic sealer. The sealer located at the facility had been leased by Coolidge but owned by GE Capital. Arch offered GE Capital $325,000 to buy out the lease. According to Leon, GE Capital expected Arch to pay $500,000 for the used equipment—as much as the cost of purchasing a new one. When Leon refused to pay the inflated price, the company argued that he owed them the money anyway simply because the equipment was housed in the building he bought. Leon countered the attack by maintaining that he bought only certain Coolidge assets and not the robotic sealer.

As Leon knew a new machine would cost the same or possibly less than the used machine, he purchased a new one. After the new machine arrived, Leon called GE Capital and told them to come and take out their equipment. "GE Capital thought they were holding all the cards," Leon said. "They thought they had four aces and I had a pair of twos. They thought I was bluffing. You do not bluff with people unless you know what is in their hand. That whole experience was actually a lot of fun."[5]

Joseph Devine, an attorney who conducted business with Leon over the years, said, "Leon is a pragmatic person. He likes to get to where he wants to go in a straight line. He's practical-minded and a tough negotiator. He is willing to walk away from a deal if he perceives that he is either being taken advantage of or he doesn't feel that it will work for the company. He'll walk. But, he's also very fair. I think that also goes toward how he treats the company's employees. He treats everyone very fairly."[6]

Enhancing the Team

Throughout this time, Leon and Bobby continued adding quality members to their growing staff. Cliff Monroe came to Arch in 2000 with extensive glass experience. Jeff Wareham, a mutual acquaintance of Monroe and Leon's, brought the two together. Wareham, who worked at HHH Architectural Tempering Systems, Inc., and knew of Monroe's expertise, suggested that he call the Silversteins about a job. Monroe explained:

> *I came down and had an interview with Bobby and Leon, and it has been nothing but an excellent exercise ever since. Arch's reputation was good. It was a growing company with a lot of things going on. It was expanding at a rapid pace.*
>
> *It wasn't my opinion, but the word on the street at the time was that it was a tough organization—tough to work for [Arch], that [it] was a low-price supplier in the market and worked [its] people to death. When I started working there, I found just the opposite. Arch was growing fast, but it was definitely a very good organization to work with. They try to maintain pricing and garner market share just by the quality of service to the customer.*
>
> *The Silverstein family is fair, honest, and they're extremely patient. They do what they say they're going to do. ... They are a very straight-shooting group of people. I think being honest, open, and having integrity is kind of rare today, and working for people [who] have those attributes is what I think is one of the best by-products of working for Arch.*[7]

Because of Monroe's breadth of experience, he would quickly become Arch's technical expert on

Arch Aluminum & Glass' industry roots date back to 1898 when Harris Perilstein first entered the glass business. Many members of his extended family would later embark on careers in the glass industry, with several relatives working side by side at Arch. From left: Kevin Aubuchon, Steven Perilstein, Leon Silverstein, Bobby Silverstein, Ricky Silverstein, Marcie Perilstein, and Max Perilstein.

glass production and quality control. Monroe serves on the boards of national industry organizations, including the Safety Glazing Certification Council, Glazing International Certification Council, Insulating Glass Certification Council, and Glass Association of North America. These organizations are responsible for setting standards and certification requirements within the glass industry, and Monroe represents Arch at industry organization meetings. Although it is a national company, Arch is considered midsize in comparison to the giant glass manufacturers. As a member of these boards, Monroe stands up for the needs of small and midsize glass manufacturers.

A Cousin Joins Arch

In October 2001, Steven Perilstein, Leon's second cousin, joined the Arch team. Steven's father, Harris, like Bobby, worked at Chromalloy for a time, but left in 1977 to start his own glass business. He opened Perilstein Distributing Corporation (PDC) in Pittsburgh. "I owe a lot of my success at PDC to Leon and Bobby because they did not compete with me in my territories, and they gave me the opportunity to get into the aluminum business," Steven said.[8] To do this, Bobby and Leon allowed the Perilsteins to open an Arch franchise called Arch Aluminum of Pittsburgh.

By 1999, Steven Perilstein, along with his father, brother, and sister, had grown their company to a $50-million-a-year endeavor. Rather than lose their independence by being forced to merge or sell, they decided to roll up their company with a new group called United Glass Corporation (UGC). "Competition was getting tougher and the independents were not going to make it," Steven explained.[9]

Customer Service Advantage

As general manager of Tropical Glass, Gene Lomando reflected on why Tropical Glass has remained a loyal customer to the Silversteins for the past 35 years:

What made Arch successful is that its employees have personal contact with the customer. To give you one perfect example, you have Arch Aluminum & Glass, a multimillion-dollar business, which has a live person answering the phone. They have thousands of calls coming in every week and every call is … personally answered by a live person, not voicemail, not putting it into a tunnel, none of that stuff.

If I had a problem right now, I could get on the phone and ask for Bobby, and that man would get on the phone. If I said I want Leon or I want Ricky, they'll get on the phone, too. The advantage they provide is personal service.

We've been very, very close over the years. We have had our ups and downs … but when we had times of trouble, the Silversteins stuck by us on a handshake or on a phone conversation. Imagine, a phone call and a handshake, and everyone got paid. We all went forward, and we've been making money together for all those years.[1]

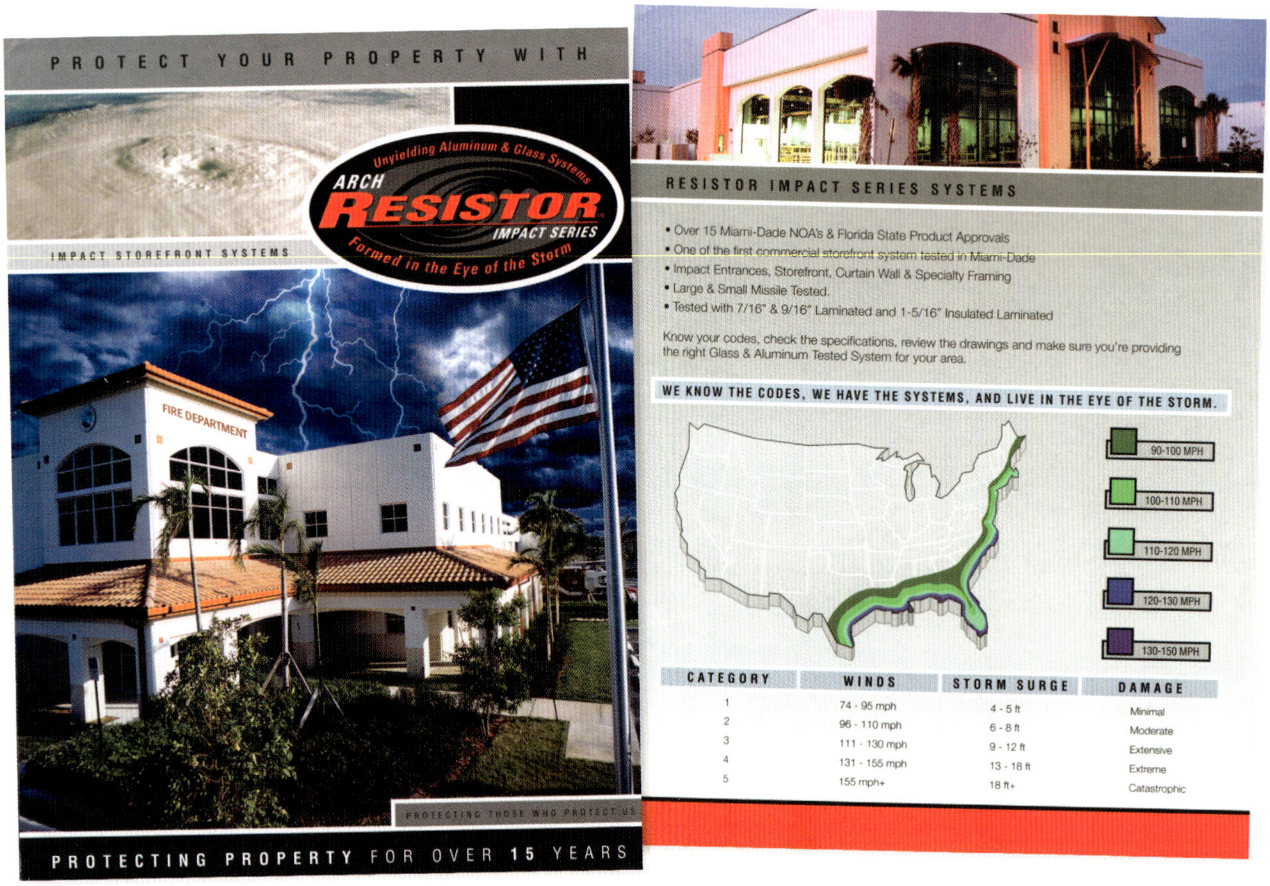

This brochure features the Arch Resistor Impact Series of impact-resistant entrances, storefronts, curtain walls, and specialty framing that have been tested and meet the International Building Code, Miami-Dade County Building Code, and Florida Building Code requirements for hurricane wind-borne debris missile resistance.

However, as part of the deal, the Perilsteins could no longer operate Arch Aluminum of Pittsburgh. The right to use the Arch name and sell Arch products had cost the Perilsteins $1 a year, and looking back, Steven said, "We should have merged with Arch Aluminum back then. But unfortunately, I had partners who would have never agreed to that."[10]

During the time of the merger, Steven's role at UGC involved working with its president and selecting other businesses to participate in the endeavor. "We wanted to make UGC into a half-billion-dollar company in a five-year time period. In the first two years, we never purchased one company. Purchasing companies required funding, and none was available," Steven recalled.[11]

According to Steven, UGC had a complicated organizational structure that resulted in a lack of communication and follow-through in business dealings. Given these working conditions, Steven decided it was time to return to Pittsburgh and the company his family had founded. Upon his return, he explained, "My right-hand man had taken over all my responsibilities."[12] There was nothing for Steven to do but act as a figurehead.

In May 2001, Steven decided that he wanted to sever his current business relationship with UGC. He hired an attorney to help him get out of his contract. Steven explained:

> *I only worked two years of a five-year contract. I went and fought that through mediation. They felt I was washed up in the industry. No one knew the real relationship that Leon and I had. We never competed. There is no way people thought I would go to work at Arch.*

Steven Perilstein and his son, Josh, who is the fifth generation of Perilsteins in the glass business, work together at Arch Deco Glass in Columbus.

UGC representatives said they would continue to pay me my salary for three more years as long as I did not go into the glass industry. If I decided to go into the glass industry, I would have to give them two weeks' notice, and I would lose my pay. They figured I was no longer a threat and that no one would hire me and pay me what I needed to be paid.[13]

During the negotiations, Steven told the group he needed to call his wife and then he would make a final decision. He went out to the parking lot and called his wife, but then he called Leon. Steven explained:

I said to Leon, "Are you ready for me to join you?" Leon asked me, "What did you get? A year?" I said, "No, two weeks." And I told him about the rest of the deal. Leon said, "Go ask for more money to pay for legal fees."

I went back inside and told them my wife said okay to the deal, but we needed more money for legal fees. They agreed to give me more money and then cut me a check. The deal was signed. I turned to them and said, "I'm giving you my two-weeks' notice."

Leon and I have always—even before I joined Arch—had a great relationship, but everyone in the

More to Learn

DAN HANEY CAME TO WORK AT ARCH in 1987 to help set up the first tempering plant in Orlando. Later, he became the national director of production for all Arch operations. In this capacity, Haney oversees the setup of new facilities, from ordering equipment to hiring personnel and helping to manage the new operations. In regard to his opinion of the Silversteins and how they run Arch, Haney commented:

They are genuinely good, hard-working people. It is why they are a union-free shop.

I learned a lot from Leon, as young as he was. He really came into this market and turned it around. I knew the mechanics of tempering glass, but Leon taught me a lot about doing business. Over the years, I was amazed at some of the things he had the guts to do. He was always ahead of the rest.

For example, in this business, you have to have the right computer software. It is the heart of the operation. We were dealing with a company that wrote the software for most of the glass industry. When Leon was expanding, each new plant needed software. He felt this company was taking advantage of him by taking stock-shelf material and selling it to him as a new custom product. He told me one day, "I've just had it with these people. I am going to write my own software." I told him, "You'll never be able to do that." And, lo and behold, with the help from his IT team, he has developed some of the best software in the industry. When he pulled that off, I was very impressed. Once you start doing things of that magnitude, people look at you in a different light. He was a real player in the industry.[1]

industry thought we hated each other. At trade shows, Leon would pick on me for wearing a suit and tie and make fun of my clothing allowance. I was white-collar in my old life, but now I'm blue-collar. I've learned. I wish I had learned these things from Leon and Bobby five or 10 years ago. They taught me a lot in the short time I've been with them. If I could be like Bobby and have the respect that Bobby has from the industry and from his workers, then I can honestly say my life has been a success.

When I started at Arch, I told Leon, "I want to be a vice president. I want this, I want that." Leon said, "No. You have to earn it. You earn the respect of the people, the respect of the branch managers. You do it the natural way."[14]

Two Families Joined

Because of his skills as a salesperson, Steven would take on overseeing the sales operations at all of Arch's branches. He commented:

Leon [and Ricky] will send me to places [they feel] need some help from a sales or from a people standpoint. It could be a new operation or a facility that's running a bit slower than it should be. I'll work with the branch managers and travel with their salespeople.

The Silversteins have given me an unbelievable opportunity, and now my brother and my sister work for them. It is like the family business all over again. Bobby has been like a father figure to me. I get along great with the whole family. It has been a great ride for me. I would never consider doing anything again without Arch.[15]

With the closure of Arch Aluminum of Pittsburgh, the Pittsburgh area became open territory for the Silversteins. They decided to open a plant in that region, and Arch set its sights on Akron. "We wanted a location to compete with the company that bought out PDC," Leon said. "We just went in and built a building. We talked to the economic development people. They helped us find land, and we bought in a city-owned industrial park in Akron."[16]

The Akron facility was planned as the largest greenfield plant to date. A state-of-the-art plant, it would take nine months to build and six more months for the ordered equipment to arrive. From its opening day, the plant was a huge success. "The funny part of the story is that now my old company is probably doing half of what it was doing in sales when I was there," said Steven. "The Arch branch in Akron—the one Arch opened to compete in Pittsburgh with UGC—is one of the most profitable branches we have, and the Arch branch manager used to work for me at Perilstein Glass Distribution. We have had more than 35 people join us at Arch from my old company."[17]

Steven's brother, Max Perilstein, also had worked in the Perilstein family business. But because of an employee contract, Max could not leave the company when it was sold to UGC. When his contract ended in 2002, he decided to join his brother and cousins at Arch Aluminum & Glass. Max explained:

Having the opportunity to work with everybody at Arch was a no-brainer, and Bobby is one of the classiest, nicest people this industry has ever seen. A lot of people don't know Bobby very well, but peo-

Twenty-year veteran employee Greg Wells (right) and Jeffrey McCalla stand in front of the autoclave-laminating machine used to manufacture glass. *(Photo courtesy of IDS Photography.)*

THUG FOR A DAY

AN ARCH COMPETITOR OFFERED STOCK to some of its employees. Through a fluke, an employee of this competitor offered to sell his stock to Leon, who jumped at the chance and became a shareholder in one of his competitor's businesses.

After the purchase, Leon requested annual reports and other information normally sent to shareholders, including a certificate of stock ownership. Only after involving an attorney did Leon finally receive the stock certificate. He framed the certificate and proudly displayed it on the wall in his office.

However, despite numerous requests, Leon did not receive any annual reports. He asked Mark Kuchenrither to pay a visit to the business and secure the reports. Kuchenrither accepted the mission. He recalled:

> *I grew my beard out for three days and wore blue jeans. I tried to make myself look as thuggish as possible. But, I was going to act like a gentleman. I was planning on sitting there all day to get the information. I wasn't going to tell them who I was. I was just going to sit there. When I got there, I was all fired up trying to find the entrance to the building when a guy pulls up next to me and gets out of his car. He was about 3 inches taller than I was. [Kuchenrither stands over 6 feet tall.]*

CFO Mark Kuchenrither looks nothing like the thug he tried to portray when attempting to retrieve annual reports from a competitor. *(Photo courtesy of IDS Photography.)*

The man identified himself as the owner's son and asked about Kuchenrither's intentions. Kuchenrither made his request and showed the man a letter from Arch's attorney. At that moment, the man opened the trunk of his car. Kuchenrither explained, "I thought I was in big trouble." But the owner's son only pulled a jacket from the trunk to retrieve his cell phone. He called his father who told him to give Kuchenrither their attorney's phone number. Kuchenrither continued, "I didn't leave. I called their attorney from my cell phone and asked where his office was. I wanted to go pick up the annual report information. I told him I had all day and that I was willing to come down and wait in his office. The attorney got nervous and said it wasn't necessary." Kuchenrither returned to Florida empty-handed, but a few days later, Leon received the annual reports through the mail.

"This company happened to be one of the companies involved in the Perilstein merger. They had treated Max and Steven very poorly. So this was Leon's way of getting back at them. I enjoyed being a thug for a day," Kuchenrither commented.

For the record, Leon eventually sold the competitor's stock, but not until he managed to make a profit on it.[1]

Above and right: Carlos Claros works as a skilled glass cutter at Arch's Tamarac, Florida, facility. He carefully positions the glass on the cutting board and uses the cutting blade to craft the glass to exact specifications. *(Photos courtesy of IDS Photography.)*

ple who do know him, people who have had the honor of working with him, respect him beyond belief. He is just a class act all the way around.[18]

The Advertising Game

When Max joined Arch, he focused his attention on the marketing side of the business. Having honed his skills in this area while working at PDC, he recognized the value and importance of advertising. But, until Max arrived, Arch did not place great emphasis on it. Max believed that advertising provides benefits well beyond the return on investment.

Shortly after joining Arch, Max purchased full-page advertising space in the major glass industry magazine *U.S. Glass.* "Arch needed to become a recognizable brand name," Max said. "We need to keep our name out there, front and center, because perception is reality. That came to legitimatize us. People came to expect it. You're not going to pick up a lot of sales from advertising, but you get exposure in the public eye, and that's what you want."[19]

According to Max, two things occurred as a result of the advertising. First, people from areas outside of Arch's local reach began making inquiries about Arch's products and services. Second, employees from competitors took notice of Arch's use of advertising that publicized and recognized its branch managers. Comments made to Max indicated that these outsiders were impressed with the respect shown to Arch employees.

Besides placing advertisements, Max developed a large marketing campaign strategy for the company, which focused on training individual branch managers and promoting what each branch had to offer. In addition, Max, along with his brother, Steven, improved upon the annual national sales meeting held for Arch employees scattered across the country. Prior to Max and Steven's involvement, only 15 to 20 people were invited to attend. However, under the Perilsteins' direction, the list of invitees grew to include sales managers, branch managers, plant managers, a host of other employees, and vendors. Steven and Max wanted to involve a large number of Arch employees to generate excitement, share ideas, troubleshoot issues, and benefit the organization as a whole.

Using sports teams as the theme for the sales meeting, Max assigned people to various teams and gave each team tasks to perform. Not only were the programs informative, they were fun. One year, Max and Steven organized a sales event that centered around ESPN sports. Leon and Steven presented a "Did you know?" fact session, which illustrated how sales could increase through specific yet simple actions and methodology.

Advertising also came into play when Arch wanted to spread the word that it manufactured the best hurricane material in the industry. Despite the fact that few major storms had struck South Florida since Hurricane Andrew, Leon told Max that he wanted to be ahead of the game. Max immediately launched a marketing campaign that continued for two years. By 2004, when four major hurricanes hit the South and Gulf Coasts of Florida, consumers and the industry were well informed regarding Arch's hurricane-resistant product line.

In addition, Arch branded its hurricane impact material with the name *Resistor*. "Everyone else uses [brand names] like *hurricane stop* or *storm guard*. We wanted to be a little different. It's a play on words. Resistor suggests that our material resists the worst that Mother Nature can bring," said Max.[20] Max explained how Arch branded itself to enhance its visibility and remain a forerunner in the glass and aluminum industry:

> *We decided to brand ourselves to make Arch stand out. We created a brand around the fact that we are the only company that can supply both the aluminum and glass nationwide. Often, architects want single-source responsibility because it is easier to deal with one party. So we coined the term and trademarked the phrase "Arch is Your True Single Source." To this day, Arch is still the only single source nationwide.*[21]

The Supreme Shower Door Glass Line is an Arch branded product line as well. Instead of offering just one or two shower door choices, as most suppliers do, Arch offers nine very distinct styles to give consumers a variety of choices.

A Busy Year Unfolds

Meanwhile, the plant in Bridgeport, New Jersey, which had opened in 1999, continued to underperform. Leon and Ricky believed in the potential of the Bridgeport market, but they felt that the business would benefit from key personnel additions, including a plant manager, an inside salesperson, and a local salesperson, so he hired like-minded salespeople, those who grasped the Arch business model, which helped to turn the facility around.[22]

Around the same time, in response to the success of Arch's mirror plants in Fort Pierce and Iowa, Arch focused on capturing the market on the West Coast and turned its sights toward California. Word got around that a mirror company called Bay Mirror of California was failing. "Bay Mirror had a wonderful reputation for having the best mirrors on the West Coast for years and years," Leon said. "But the owner was getting old and tired. We went there, and the mirrors were great, but the plant was a pigsty. I would tell people, 'I went out there and put on my rose-colored glasses. I fil-

Sister Assistance

NATALIE REED, BOBBY SILVERSTEIN'S younger sister, has worked in the glass business with Bobby and their father, Sydney Silverstein, since the early days at Architectural Glass Products. She attributes the family's success to Sydney and what she and her siblings learned from him. She explained, "He was very successful, and he set a wonderful example for us. He trained his children well."

When Bobby left Chromalloy to launch his own business, he asked Natalie to come to work at Arch. She agreed:

> *I started with Bobby in 1980, and I went into human resources. I was basically his assistant because the company was so small at the time. Later, I traveled a lot for him to the different locations, setting up each facility's insurance and human resource programs. I loved that. When I was down there at the Pompano facility, I was doing the administrative portion of human resources. Now that I'm up here in Fort Pierce, I do it all. I do new hiring, insurance, workmen's compensation; I do whatever needs to be done because I'm an old lady now. But I have a great boss, too. Tony Lawson is my boss, and I love working with him.*[1]

www.arch.amarlite.com
ARCH Aluminum & Glass
Mail
Favorites
Larger
Smaller
Preferences
TRUE SINGLE SOURCE
Visit www.arch.amarlite.com and learn about the capabilities of Arch Aluminum & Glass Co., Inc. manufacturers of Amarlite Architectural Products. Arch can satisfy virtually any architectural glass and aluminum requirements.
From the distribution of Flat Glass...
To the fabrication of Float, Mirrors and Laminated Products...
To the manufacturing of standard Doors & Framing...
To multiple spanning – continuous beam Curtain Walls...
To the durable Powder Coat finish...
Arch Aluminum & Glass is your "True" Single Source supplier.
Locations Nationwide
1-800-432-8132
Curtain Wall Systems
Entrances & Storefronts
The Pressure's On...
Are your local codes changing?
Will you be compliant?
Find out at our web site!
Aluminum & Glass Co., Inc.
Manufacturers of Amarlite® Architectural Products
Circle Reader Card #02

tered out all the bad stuff. I saw a good line, a good product, and some good people.' The aging owner had had a few bad years. He did not want to put any money into the plant nor did he want to grow the company."[23] So Leon made an offer and bought out the owner of Bay Mirror, renaming it Arch Mirror West.

Once again, Arch utilized the expertise of Numa and Mary Handi. They flew out to California and spent more than a year working on improving the plant's operations. Despite their best efforts, the plant did not perform as well as expected due in part to stiff competition from China's less expensive mirror imports. It would take another five years to make the plant profitable through aggressive marketing, sales efforts, and price reductions. Forced to reduce prices because of the competition, Leon and Bobby insisted on maintaining the high quality of their products. "People want top quality but are not willing to pay for it," Leon observed. "Market pricing is dependent upon region and freight costs."[24]

Throughout 2001, Arch continued to expand its influence in the industry and across the United States. In July 2001, *U.S. Glass* magazine featured a story on the industry's most influential companies, along with a list of company heads, or "power rangers," as the magazine called them. Arch Aluminum & Glass and Bobby Silverstein graced those pages. Despite the national exposure, and the growing recognition that Arch received, Bobby and Leon continued to foster the small mom-and-pop culture so critical to the success of the company.

Setting Up Shop

Meanwhile, an existing plant in Dallas, Texas, called D & S Glass, declared bankruptcy. The bank intended to sell all the company's assets at an absolute auction, meaning everything—regardless of price—would be sold. Recognizing an opportunity for Arch, Leon went down to Texas to look at the assets. He immediately saw the potential for Arch's entrance into tempered glass in that region. If the plant did not work out, Leon reasoned, he could shut it down and move the equipment elsewhere.

Arch had actually considered buying D & S in 1998, so Leon was familiar with the financials of the company and recognized a good buy. He contacted the landlord and negotiated a deal to lease the building before the bank put the assets up for auction.

With the lease deal sealed, Leon went to the auction with $500,000 in cashier's checks. One by one, he bought each piece of equipment, and by the day's end, Arch would own everything needed for its Dallas plant. "The glass industry is a small industry, so within an hour of the auction, everyone knew we were coming to Dallas. Many of the employees of D & S Glass were at the auction, anxious to know who was going to buy the plant. This was big news

Opposite: This July 2000 advertisement emphasizes Arch's ability to serve as the true single-source supplier for every architectural glass and aluminum need. It also highlights Arch's Web site as a multifunctional tool.

Right: Jeff Ziesche followed Steven Perilstein from Arch Aluminum of Pittsburgh to join Arch and focus on increasing sales and service in its aluminum division. *(Photo courtesy of IDS Photography.)*

MEMORY LANE

MAX PERILSTEIN MET HIS WIFE, Beth, while he was working for Perilstein Distribution Corporation. Max recalled:

> *My wife is actually from Detroit. She worked for a company that was a customer of ours. She would call and place orders. I would flirt with her on the phone. I went up to Detroit, and I got to meet her. I fell in love. Somehow or other, she fell in love with me. I encouraged her to move to Pittsburgh, and we got married.*
>
> *Then, when we had an opportunity to acquire a company in Detroit, Beth and I moved back to Michigan just to run the company. The company that Beth had worked for prior to our marriage was so mad that she'd left them to marry me that they never bought from us ever again. At the time, they were also buying from Arch. Then, when they found out that I went over to work at Arch, they stopped buying from Arch. They thought I had stolen her away from them. I didn't steal her away to work for us. I married her!*[1]

since no one knew we were coming [before this]," Leon commented.[25] It took 90 days to clean up the plant and refurbish the equipment, using what Leon calls the "swat team" of workers from Arch's machine shop in Orlando.

Arch's swat team ensures the proper functioning of equipment and pays attention to every detail associated with the plant's opening. The team spends two solid weeks at the new facility and then returns to work at the Orlando shop for two weeks. Then the team heads back to the new facility for another two weeks, and so on, until everything is in order and ready for opening.

Shawn Myers, who joined Arch in 1985, works on the swat team out of the Orlando plant. "After 20 years of working at the Orlando facility, you get to know people, and there's always a new thing happening," Myers said. "It's always an experience. I mean, with us going to different plants and traveling around, you just get a lot of knowledge."[26]

To get the Dallas plant fully operational, the Arch swat team installed a new telephone system, painted the interior and exterior of the building, reorganized the location of the equipment, installed a new cutting table, and built new racks. In addition, Leon purchased a new fleet of trucks to service the area. Haney, who assisted with the setup of many Arch plants, including this one, became the interim plant manager and senior manufacturing expert there.

After the opening, Arch already had its sights set on another facility. An Arch Aluminum & Glass competitor, American Architectural Products, had purchased several smaller companies, including American Glassmith of Columbus, Ohio. But in 2002, American Architectural Products declared bankruptcy. An employee at Long Point Capital learned of the bankruptcy and suggested to Leon that he explore the possibility of purchasing some of the company's assets, including American Glassmith.

American Glassmith distributed decorative glass, a product line as yet untapped by Arch. Upon careful examination of their financials, the family discovered that although the business was losing money, only one product was to blame: leaded glass for entrance doors. The company priced the leaded glass as a custom product. But the product competed in the marketplace with non-custom products, which sold at non-custom prices. That product alone, management reasoned, caused the entire business to fail. Management believed that by eliminating the leaded glass product line, Arch could make the plant thrive. According to Don Hyde, senior accounting executive at Arch, "[The family] has the uncanny ability to walk through a plant and see its value immediately."[27]

David Williams, former president of American Glassmith, who now heads the decorative glass division for Arch, said:

> *We always knew we had a good product line and that if we could eliminate that one product, we could have a profitable business. We were a good fit with Arch. Culturally, we were very similar. We closed as American Glassmith on Friday and reopened on Monday as Arch. And [Arch] pretty much left us alone, and almost immediately, we were profitable.*[28]

Getting the Aluminum Side Geared Up

For many years, Arch primarily focused its expansion on the glass side of the business. To build up the aluminum side of the company, Arch hired Jeff Ziesche in 2002. Ziesche possessed extensive aluminum experience from his involvement in the industry since 1987. He ran the aluminum division at Arch Aluminum of Pittsburgh for Steven Perilstein until he left to come to work at Arch.

Ziesche knew of Bobby and Leon from their relationship with the Perilsteins and Arch Aluminum of Pittsburgh. "When Steven was leaving Pittsburgh to go to Arch, Leon called me and said, 'You know, if things don't work out up there, give us a call. Keep in touch.' So we did keep in touch," Ziesche recalled. "Then, more things transpired in Pittsburgh, and I decided I was ready for a change. Leon and I started talking, and he invited me down. Leon said, 'We know your experience, Jeff. Why don't you bring your wife down and look around? If you feel comfortable, we'll make a deal.' "[29]

After visiting Florida and spending time with the Silversteins, Ziesche decided to move his family and join Arch. "They're a family business, which I like. It wasn't like I was being interviewed when I went down there," Ziesche said. "It was more like we were just talking. I remember Bobby making a comment to me. He said, 'You know us, Jeff. You know what we're all about.' "[30] Ziesche agreed and got to work. He explained:

> *My goal when I came in was to really build the aluminum division up. My goal was for aluminum to be 35 percent of overall sales. It is a lofty goal because any time we make a new acquisition, there's more glass being sold. Right now we're at 25 percent overall sales in aluminum.*
>
> *Arch is a major corporation, but we're also a family business. We set up our branches and sell our products like a small family business. A typical branch services a 200-mile radius. We deliver using our own trucks and service that area for local customers. We present ourselves as a major corporation, but with that local flair.*[31]

Now based in Florida, Ziesche travels frequently to different branches across the country to maintain communications, solve problems, and improve business on the aluminum side.

Tightening Operations

Changes continued to take hold at Arch during this time. The Arch Board of Directors held regular meetings; reviewed financials, capital expenditures, acquisitions, and pricing; and closely monitored every aspect of Arch Aluminum & Glass. By the middle of 2002, the board had approved several acquisitions, including Arch Mirror West (California),

Arch Aluminum & Glass received the 2003 Manufacturer of the Year award from the South Florida Manufacturers Association for its ongoing commitment to the manufacturing industry.

Arch Portland, Arch Milwaukee, Arch Dallas, and Arch Akron.

The board also instituted improvements on existing facilities. They decided to make management changes at the California mirror plant and the Bridgeport facility to address personnel issues. Because the Portland plant, a tempering facility capable of creating oversized glass, continued to underperform despite strategic efforts to boost it, the board voted to stop taking new orders between 2001 and 2002 in preparation for the plant's closure. "The lease on the building was up," Leon explained. "It was also an undersized building, which made production inefficient. In addition, the plant was next to railroad tracks, so every time the train came by, it would kick up dust, blow it into the blowers, and we would end up with a lot of rejected glass. We also failed to put the proper people resources into place. Another lesson we learned: Without proper management ... success [is] very hard."[32]

A slow regional market and the availability of cheaper imports arriving from Canada contributed to Portland's underperformance. With the plant's impending closure, Leon and the board decided to take the equipment from that facility and move it to the newly acquired facility in Southern California to increase its capacity.

Continuing to plan for growth, Arch intended to open another laminating facility in the southeast region of the country with the possibility of adding an extrusion plant.[33]

Expanding in Manufacturing

Binnings Pan American (BPA), a Miami-based window company, went into bankruptcy in 2003. Leon commented: "The salesman for that press was named Mike Rodriguez. Right before the BPA shut down, we heard they were having trouble. We also found out from Rodriguez that the extrusion press at BPA was leased. We contacted the leasing company and made arrangements to buy the press. We also made a deal with the building landlord to lease the space where the extrusion press was located. To move the press would have cost us up to $500,000."[34]

In return for Rodriguez's tip, Leon agreed to hire him to run the extrusion press for Arch. Rodriguez wanted a change from selling to working on the manufacturing side of the business. "We knew that Mike's

Below and right: The process of extruding aluminum has been compared to squeezing toothpaste from a tube so that the paste remains in the shape of the tube opening. A piece of aluminum, called a billet, is heated to 800 to 1,000 degrees Fahrenheit and forced under pressure through a die of the desired shape. The aluminum exits the press and is cooled by a series of fans on the run-out and cooling table. *(Photos courtesy of IDS Photography.)*

In their home office, Beani Silverstein stands by her husband, Bobby, as she has through their marriage of almost 50 years. *(Photo courtesy of IDS Photography.)*

father had been in the extrusion business for more than 30 years. We knew if Mike didn't know how to do something, he could always ask his father," Leon said.[35]

An auction was held to liquidate BPA's remaining assets. Leon attended the auction and put a sign on the press that said "Not for Auction," as he had purchased it already. According to Leon, "The owners of the window company thought they could sell the press at auction, unaware of the fact that I had already bought it from the leasing company. Even though they didn't own the press, they thought they could sell it at auction for a certain price and then make a deal with the leasing company to purchase it at a lower price, thus pocketing the difference."[36]

Purchasing the extrusion press proved extremely valuable for Arch because now the company could manufacture its own aluminum—adding still more value to its hallmark one-stop shop. Within one year, the extrusion plant would run 24 hours a day, six days a week to meet customer demand.

Stepping Up to the Plate

During the 1990s, Ricky Silverstein spent most of his time working in sales at Arch. By 2002, he oversaw all the branch operations in the southeast region of the United States. The position of chief operations officer had been filled by an outsider for a short time only, but for a variety of reasons, the match did not work. Ready to expand his involvement in the company, Ricky stepped into that role. "I'm kind of mellow. I can handle stress well. I had a good rapport with all the guys in the back. I never *tell* them what to do. I *ask* them to do things, and I think that's why they've always done what I ask," Ricky explained.[37]

Donna, Leon and Ricky's sister who would soon join Arch full-time and work with her husband Kevin Aubuchon in the IT department, commented, "People in the back were looking for a new leader."[38]

Logistics Follow Operations

Because of Arch's unprecedented expansion, Leon and the board of directors recognized the need to hire an executive who could effectively manage all the logistics associated with a fast-growing organization. They chose Daniel Wagner, who had

Dan Wagner joined Arch in 2002 to manage logistics and purchasing for the company. He was attracted to Arch, like so many before him, because of the family-oriented company culture. *(Photo courtesy of IDS Photography.)*

extensive experience managing logistics and purchasing for large corporations. "I knew Arch was a family business. The recruiter told me it was a strong company moving ahead. So, I came down to interview," Wagner recalled. "Bobby approached me that day and said, 'Dan, you have a real challenge ahead.' On my way home, I thought, 'How bad can it be?' But when I walked in the door, he was pretty much right."[39] Wagner could not have anticipated the lack of structure or coordination between branches that existed within Arch at the time.

Kevin and his team had already begun the process of implementing an enterprise resource planning (ERP) system that would enable Arch to centralize and computerize all purchasing and inventory information for the company. Still, according to Wagner, "The staff was not as strong as it should have been. There was no coordination from the corporate side. There was no one to do corporate pricing, which would greatly reduce operating costs."[40] Wagner reviewed accounts payable to understand the process. "It was a manual process. I had to understand what the commodities were and who the suppliers were before I could go out and talk to the vendors about corporate contracts. Because each branch handled its own ordering, I spent hours and hours—seven days a week of analysis."[41]

Nonstop Growth

In 2003, Leon recognized an opportunity to expand into California. The Silversteins had considered purchasing a company called Trident Industries a few years back, but they found the asking price too high at the time. Since then, the owner of the company had died and the owner's son and wife lowered the price considerably because they wanted to sell the business."We ended up buying the company rather than just the assets. The family was very nice," Leon said.[42]

But it would take several years to make the Trident facility profitable. "The personnel were very set in their ways. We would go to the plant and explain how product was to be made and delivered," Leon explained. "They would claim that they understood, but as soon as the Arch team would leave, the personnel would go back to the way it had always been done in the past. We were finally forced to hire a young aggressive manager who understood how the plant needed to operate."[43]

Kuchenrither on Board

As Arch's business continued to expand, its needs outgrew the ability of its chief financial officer (CFO). Because of MASCO's relationship with Long Point Capital, Leon contracted MASCO's services to conduct a recruitment search for a new Arch CFO. In August 2003, the human resource representative pursued Mark Kuchenrither for the position. Kuchenrither had experience as a CFO for companies in both the automotive and electrical industries. After two interviews with the human resource people at MASCO, he flew down to meet with Leon and Bobby. Kuchenrither said:

> *Leon and I hit it off right away. I saw in him a guy who was my age—young and very aggressive. His ideas about risk taking fell in line with my thoughts and ideas about risk taking. And I felt like I could really work for him. I thought that he would be a great*

mentor to me, but at the same time, I felt like I could really help him take the company to the next level.

I think I also hit it off with Bobby right away. I just really enjoy [him]. He reminds me a lot of my dad, in his value system and what he thinks is important. He is honest. They are all honest. Leon, Bobby, and Ricky are as honest as they come, and I really value that. They are good people. They don't let success create a barrier between the employees or people in general. I could sense that these were the kind of people I could work for.[44]

Leon and Bobby were impressed with Kuchenrither, too, and decided to make him an offer. Kuchenrither explained:

The great thing about this story and about the relationship is normally, with these types of jobs, you have an employment agreement. You get the offer on paper. You want things in writing because you always want to be able to point to the sheet of paper and say, "This is what you promised me. You need to live up to this."

So it's a good idea to have these things put down on paper. When I asked for that, Leon came back and said, "If you need it in writing, I'll give it to you in writing, but our family does things with a handshake." I remember thinking to myself, "What do I make of this fellow? Do I trust him? Or do I require an agreement in writing?" I turned to Leon and said, "Look, let me think about it. What I'd like to do is bring my wife down to make sure she's on board with the decision." So I brought her down and we actually had dinner with the whole family. They were charming.

Afterward, we went back to the hotel, and I asked my wife what she thought. She said, "I think you should do it." I went to Leon and said, "I'd like to work for you." We shook hands, and that was it.[45]

In terms of business dealings, Kuchenrither learned how to understand and value Leon's business instinct. According to Max: "Mark is a by-the-book type of CFO. He works the numbers. Leon doesn't work that way. Leon can see things and feel things. One day, Mark said to Leon, 'I need to put

The Sydney Awards

The Sydney Awards are presented to selected employees at Arch's annual national sales meeting. Since 2003, the awards, named in Sydney Silverstein's honor, recognize excellence and overall improvement in key areas, including:

- True single-source shop
- Best branch of the year
- Signature job of the year
- Newcomers of the year
- Salesperson "hustle"
- Most improved branch
- Best truck dispatch
- Best days outstanding
- Most improved days outstanding
- Best safety performance
- Safety—less than $1,000 cost for injuries[1]

In 2005, Gerald Ashley (center), longtime Arch employee and branch manager at the Youngsville, North Carolina, facility, received the Sydney Award for "Best Branch to Work With" from Ricky and Leon Silverstein.

THE MORAL OF THE STORY . . .

Bobby Silverstein has a knack for telling colorful stories that get his point across about business in a comfortable, nonconfrontational manner.

As the new CFO for Arch Aluminum & Glass, Mark Kuchenrither recalled some of the valuable lessons he has learned from Bobby Silverstein:

Bobby Silverstein has always made very sure that when we walk into Arch, we leave our egos at the door and that we don't flaunt our success in front of other people. He never lets us forget that the guy in the back who is making $10 an hour is just as critical to the success of the company as somebody sitting in my chair.

He'll come into my office and say, "I'm not telling you what to do and I'm not giving you advice, but let me tell you a story." Then he'll tell me a story and I'll say, "Okay, this is what I'm going to do." He'll say, "I'm not telling you what to do." And he'll get up and leave. And I always would get his message. Oftentimes, the stories would be about how to deal with people or manage situations. Bobby is very good about showing compassion, but he also knows when someone is taking advantage of him.

Bobby told me a story about this fellow who ran one of the branches in Florida. Financially, he had a good year, and this fellow had been instrumental in turning around the financial performance of the facility.

The branch manager had an office for himself. The facility had two other offices with a woman working in each one. One day, this manager made the two women move into one office. He knocked down the walls and expanded his office and put in, I think, a bathroom and a fancy bar. Bobby caught wind of this and went up to visit the guy. The branch manager thought that Bobby was coming in to congratulate him and give him a raise, and pat him on the back. But Bobby went in and fired him.

Bobby laughs about it because he says, to this day, he'll see this fellow from time to time, and the guy still doesn't understand why he got fired.[1]

some numbers down on what you think this acquisition might bring.' Leon said, 'Don't worry about it. I know in my gut that this is a good one. Trust me, my gut says it's good.' Mark looked at Leon and said, 'I can't take your gut to the bank,' to which Leon responded, 'Oh, yes, you can.' And then he walked out. Mark assimilated very quickly, and he takes his cues from Leon, and then goes and works up the numbers. That is why those two work so well together."[46]

Employee Wellness

Besides providing employees with business advancement opportunities within the organization, Arch actively encourages personal wellness for its workers. Every month, Arch provides a newsletter to employees that outlines the various health care plan options available, as well as information on keeping health care costs down. The newsletter also conveys clear explanations about various benefits offered to employees, including practical information on health care spending accounts, as well as useful information on common health problems.

One month, the newsletter announced a "Biggest Loser" three-month weight-loss program and contest. With more than 445 participants, Arch paid out $18,000 in incentives to employees who successfully lost weight. "The program really gave people something to talk about," Kuchenrither explained. "Employees at branches all over the country were excited about it. Some people weighed in together. Others had buddies that they would eat lunch with and offer support to. It was a real morale boost for the employees, many of whom changed their overall lifestyle as a result of the program."[47] In the future, Arch plans to establish health contests and

The Silverstein family members who work at Arch pose on the mezzanine at the aluminum facility in Tamarac, Florida. From left: Ricky Silverstein, Donna Aubuchon, Kevin Aubuchon, Bobby Silverstein, and Leon Silverstein. *(Photo courtesy of IDS Photography.)*

programs to improve blood pressure levels and cholesterol readings, to name a few.

Recapitalization

In 2004, the Silversteins decided to recapitalize to gain a larger percentage of ownership in the company. Their investment banker, Piper Jaffray, assembled and presented a prospectus about Arch to a select group of companies. Cravey, Green & Wahlen (CGW) was one of the equity investors approached. Its history of investing in distribution and fabrication companies was a good fit with Arch. "We are the oldest in the Southeast making those types of investments," said Kevin McCarthy, a partner at CGW. "We are focused on being partners with management rather than telling management what to do. That is the way we run many of our businesses. The Silversteins were looking to own half of the company. We liked the fact that the family wanted to keep the money in the business."[48]

CGW received the prospectus describing the family and the Arch Aluminum & Glass business, and, less than a week later, CGW partners Dick Cravey and McCarthy flew down to meet the Silversteins.

"We liked the nature of the business and immediately liked Leon and the family. We were impressed that Leon had taken steps outside of the family to bolster the management team," Cravey said. "A lot of what we do is chemistry-based. We pretty quickly moved from doing analysis to wanting to do business."[49] McCarthy added: "Leon had a very entrepreneurial philosophy. And our firm is similar. It became apparent very quickly in the meeting. Also, it was apparent that family members who were involved in the business were there because they contributed. But, you need to augment the team

Left: Leon Silverstein's vision to grow the company aggressively remains paramount, as does his determination to maintain Arch's homespun company culture. His family dog and trusted companion, Jesse, spends every day at the office and adds to the casual corporate feel.

Below: Jackie Audette, who joined Arch in 1999, keeps busy and smiling as the executive assistant for the entire Silverstein family—Bobby, Leon, Ricky, and Donna. She also assists Kevin Aubuchon and Mark Kuchenrither as needed.
(Photos courtesy of IDS Photography.)

A Company with Heart

Arch Aluminum & Glass supports several charitable organizations. Every year, the Silversteins sponsor a charitable event that provides equine-assisted therapy to physically and mentally challenged children and adults through Horses and the Handicapped of South Florida, Inc.

Also, Arch raised more than $15,000 a year for each of the past five years as a site sponsor for cystic fibrosis. The company first got involved with the fundraiser because an Arch employee had grandchildren who suffered from the disease. Besides supporting the Cystic Fibrosis Association, Arch gathers and delivers carloads of toys for Toys for Tots each holiday season.[1]

Marnie Silverstein, one of Leon's daughters, speaks to a crowd at the Horses and the Handicapped event sponsored by Arch. *(Photo courtesy of www.semartinphoto.com.)*

with people from the outside. That is not an easy thing for family-owned businesses to handle. Leon had done a good job handling that."[50]

Over the years, Arch expanded from primarily serving as a distribution business to one that manufactured product as well. In Cravey's opinion, because a manufacturing business depends on capital to buy the equipment, it must be run differently from a distribution business. Cravey and McCarthy saw that Leon understood this business concept and that he would take the necessary step to manage the business from more of a manufacturing perspective.

"Less than an hour into the meeting, Leon was still explaining to us why [their] business was a good business," McCarthy said. "Then Cravey interrupted him and said, 'Leon we get it. What's it going to take for us to become your partners?' From that moment, we were moving down the path to close the deal."[51]

"The culture Bobby began is a great way to run a business," Cravey added. "It is very egalitarian. We bought into the hardworking, blue-collar attitude about things 100 percent. If you go to one of the sales meetings, you can see it. It is a very flat organization in terms of how everyone gets along."[52]

The recapitalization would enable Arch to secure a line of credit, which in turn would allow the family to buy back more than 50 percent of the business.

Internship Program

The family-run business now turned to incorporating even more of the family into the company. Arch's first summer internship program was launched in 2004 as a means for some of the younger Silverstein family members to gain practical work experience before heading off to college. Leon's daughter Marnie as well as Kevin and Donna's daughters, Maddie and Rachael, and a few of their friends participated in the inaugural program managed by CFO Mark Kuchenrither. "I brought in six kids that first summer. I interviewed each one and told them they would be treated as adults and that I expected them to behave as adults. I put them to work as financial analysts. They worked on creating performance metrics for me with Excel and PowerPoint programs. I spent a great deal of time teaching them how to analyze data, how to create the drafts, and how to put the information together. Then the interns actually e-mailed their findings to

Arch started its first internship program in summer 2004. The students, many of whom were family members, gained insight into business operations at Arch. From left: Tanya Vartanian, Maddie Aubuchon, Rachael Aubuchon, Scott Stern, Marnie Silverstein, and David Rothenberg.

our branch managers. So during the course of that first summer, they started to learn the basics of how to analyze financial and operational information to run a business," Kuchenrither recalled.[53]

In addition to gaining practical work experience, the students perhaps gained something even more valuable: They learned the value of networking and nurturing business relationships. Kuchenrither noted:

> *In one case, I wrote to one of the intern's deans at her business school. I've also provided letters of recommendation for several students. The way I view it is that we may not have any of these students come back and work for us full time, but we want to invest in them because they are going to be good citizens, and we want to feel like we played a positive role in building them as productive individuals.*
>
> *Personally, I've gained tremendous satisfaction from participating in the program. As a matter of fact, at the end of the program I asked each intern for one thing: a photo of themselves. I told them, "I will frame it on the wall and when you are famous, I want to be able to point to that picture as an old man and say that you started here with me."*[54]

Tamarac Plant Addition

By 2004, Arch had outgrown its offices in Tamarac. The company decided to build an additional facility across the street from the original building. Because the laminating facility in Fort Pierce had reached capacity, the new building in Tamarac would house an autoclave for laminating. It would also provide more office space for the information technology and accounting departments. Most importantly, Leon said, "We could do a better job of servicing our South Florida customers with a laminating facility in Tamarac. We were not doing a very good job since Fort Pierce was at capacity."[55]

The two-acre parcel of land slated for the facility consisted partially of wetlands. The town building guidelines dictated that wetlands could be used for commercial purposes only if new wetland areas were created. As a result, besides building the new facility, Arch needed to help finance the wetlands project in the Tamarac Community Park. "This was a very expensive piece of property to develop. But it's nice having the two facilities next to each other," Leon explained.[56]

Concurrent with building the new Tamarac facility and managing the recapitalization from Long Point Capital to CGW, Arch acquired another facility in Denver, Colorado. American Flat Glass (AFG) announced that it would close down its Denver facility or put it on the market. Always on the lookout for favorable markets and locations, Leon saw this as a great opportunity. He immediately called AFG and negotiated a deal to purchase the assets of the Denver plant. "AFG still had 12 months left on the lease," Leon recalled. "We contacted the landlord and made a deal. We had the AFG lease ripped up and had a new lease drawn up for us."[57]

Shortly after purchasing the Denver facility, Leon heard rumors regarding the sale of Glass Tex, a company in Houston, Texas. Built in 2000, the Glass Tex plant was laid out with the help of the HHH company. According to Leon:

> *The plant looked very much like an Arch facility. As the business failed, all the investors bailed, except Jim Adams. He is one of the original founders of Piggly Wiggly. He is a very smart gentleman. He made changes and brought in his own people. He would drive over 900 miles every six weeks to check on them. By 2004, he figured he could sell it and get his money out. When we learned about the sale, we called him. Gene McKay, the branch manager there, worked for us in Kansas City several years earlier. Gene McKay talked to Jim Adams and told him we were good people.*[58]

Leon called Adams and asked him what he wanted for the business. "In response, Adams wrote a plain, simple letter stating that he wanted only as much as he had invested in the plant, and he quoted the amount," Leon said. "I asked my dad how much money I should offer him. He said, 'Don't negotiate with him at all. If you negotiate with him, you will get him upset. He is an old-timer, and he told you exactly what he wants, so offer him that amount.'"[59] Leon took Bobby's advice and offered Adams his asking price. With that, they struck a deal. "Adams is just the salt of the earth," Leon said. "The nicest man I ever met. The day we closed, Adams was there writing bonus checks to his employees."[60]

Over the years, Bobby (right) has offered Leon and others excellent advice on communicating effectively with employees, customers, potential clients, and business associates. *(Photo courtesy of IDS Photography.)*

Prior to the closing, Leon took Adams' managers, Gene and Shirley Taton, out to lunch. Shirley recalled: "Gene and I turned the plant around to where it was making a profit. Leon wanted to let us know that he was going to leave us in charge. I liked Leon. I liked the way he spoke to us. About a month later, Leon came to the plant for a day or so. He said to us, 'Keep doing what you are doing.' And then he left."[61]

It was during this time of expansion that Leon's sister, Donna, who had never worked for Arch, decided to become a full-time IT consultant. "They bought this ERP system, the new computer system," Donna said. "Kevin would come home at night, and we would discuss his work and the challenges he faced. I would give him advice. I would educate him, and he would say, 'Why don't you come work at Arch?' Mark Kuchenrither thought it was a good idea. It was my skill set they needed more than anything else, and it has turned out very nicely."[62]

More Acquisitions

In 2005, Beani's cousin and Bobby's old friend and colleague, Ray London, decided to sell his large glass distribution plant and Arch Aluminum of Dallas—companies he had opened after leaving Chromalloy in 1980—to the Silversteins. Leon explained: "[Ray] could have changed the name of his company and start positioning it as his own company or become a part of us. He sold the business to us and became a shareholder in Arch. At the time, his leases were almost up, and one of the places needed a lot of work, so we decided to combine the two Dallas places into one 150,000-square-foot building. That's a big building, a big facility, with everything."[63]

During the same time frame, one of Arch's largest customers, KraftMaid Cabinets, a national cabinet manufacturer, planned on opening a new facility in Salt Lake City, Utah, but it needed suppliers. Representatives from KraftMaid approached Dave Williams—who managed Arch's decorative glass plant—to see if Arch would open a glass plant in the Salt Lake City area to service them. Because KraftMaid buys a great deal of product from Arch, Leon agreed. In no time, plans were under way to establish an Arch facility in Salt Lake City. This plant would fabricate and distribute decorative glass and serve as a tempering facility.

As Arch's decorative glass plant in Columbus, Ohio, outgrew its available space, management decided to move the plant to a bigger facility that could also house the necessary equipment for fabrication. They found a 100,000-square-foot facility. With this fabricating plant, Arch would now boast five laminating facilities.

Arch experienced continual growth in 2006, strategically planned for by the family. During this period of nonstop expansion, Arch added a laminating facility to the Orlando, Villa Rica, and large Dallas facilities. In addition, from the outset, the Miami extrusion plant worked at full capacity. Rodriguez, the plant manager, requested a second extrusion press to handle the nonstop production and large distribution customers. Uncharacteristically, Leon asked Kuchenrither to provide financial justification for the purchase. "This was one case," Leon recalled, "that I couldn't do on feel. I had to see it on paper."[64] Kuchenrither provided the financial projections to rationalize the $8 million expenditure, and Leon approved the purchase.

A new and larger facility was needed to house the second extrusion press as the current plant's lease would expire in 2006. "We started looking for a 150,000-square-foot building in Florida," said Leon. "We looked everywhere and we were getting frustrated because we couldn't find a suitable

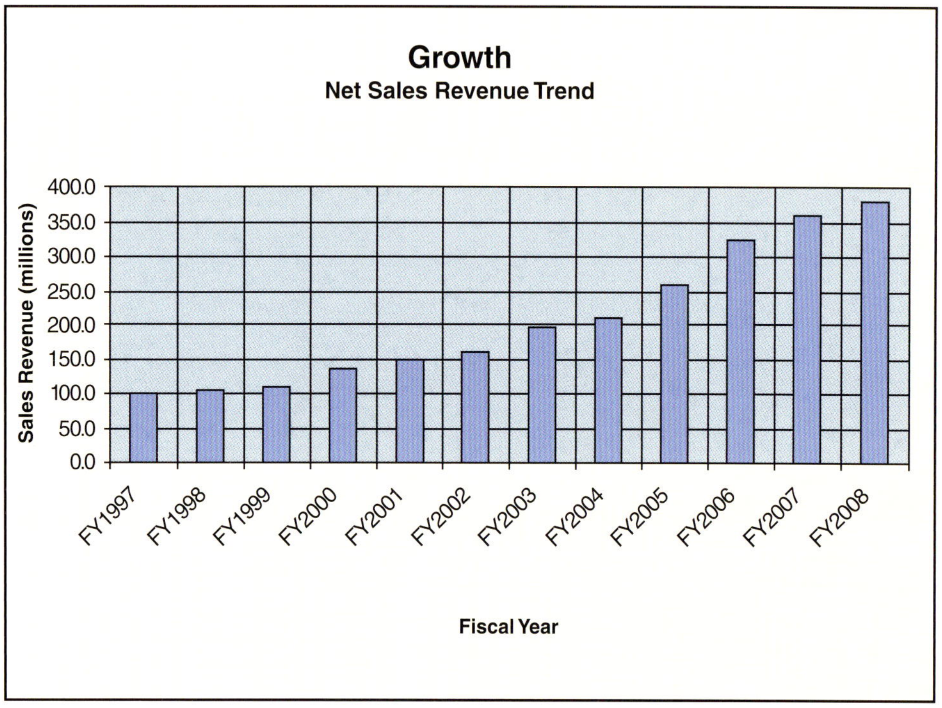

This chart illustrates the impressive and continual capital annual growth rate (CAGR) Arch has enjoyed since 1997.

STRATEGIC DIRECTION

Invest in Capacity and Equipment Technology

	1999	2000	2001	2002	2003	2004	2005	2006	2007	Projected 2008
Total										
Growth	5.2	6.5	8.1	1.8	4.4	3.4	3.8	9.8	6.6	4.0
Maintenance	1.1	2.4	2.6	2.1	1.5	1.8	2.3	3.6	3.4	3.0
Extrusion Press								6.4	1.4	1.0
Green Fields									5.0	
Total	6.3	8.9	10.7	3.9	5.9	5.2	6.1	19.8	16.4	8.0

location. The Miami landlord called us and said, 'I heard a rumor that you want to leave. Is it true?' "[65] Leon explained his situation to the landlord, and the landlord offered to make room for Arch by removing his current tenants. Leon said, "They put together a very good deal for us. Unlike the other tenants, we paid our bills. Now we didn't have to relocate. Staying in the same location presented less risk."[66]

By this time, Arch had expanded to 32 locations. Within those 32 locations, Arch has 16 traditional glass manufacturing facilities, two decorative glass manufacturing facilities, three mirror manufacturing facilities, two window and door manufacturing facilities, six distribution centers, eight laminating lines, and one extrusion facility.

When asked what he anticipates for the future, Leon said, "We'll look to expand product line and to manufacture more of our own products. Anything is possible as far as manufacturing and producing goes. In addition, our goal is to grow facilities all over the United States and beyond. We're going to really grow depending on what opportunities present themselves."[67]

With honest, hardworking staff at all levels, from the industrious people in the warehouses to the outstanding sales team and talented branch managers, Leon sees no limit to Arch's potential. Leon noted:

Arch continues to serve as the low-cost provider in the market and to invest in technology, upgrade equipment, and grow the business. This chart illustrates the large investment the Silversteins make each year in capacity and equipment technology. The numbers shown represent dollars in millions.

> *The only thing that limits us is the lack of skilled people. As long as we have good people that can run and work at these facilities, we won't stop growing.*[68]

Since its humble beginnings in 1978, Arch has grown from a company with a single truck and three employees to become one of North America's largest architectural glass and aluminum fabricators and suppliers. This family-owned business now operates facilities throughout the United States, employs more than 2,000 people, and boasts sales of more than $300 million.

Arch bases its business philosophy on creating long-standing relationships built on trust and fair play and on providing its customers with high-quality products at competitive prices. Arch's unique family culture and business values will remain its hallmark by encouraging its limitless potential for future success.

APPENDIX

ARCH ALUMINUM & GLASS' SIGNATURE PROJECTS

Signature projects are those jobs that only a limited number of companies have the capability to fabricate. The jobs excite Arch because they are unique and offer creative challenges.

—Leon Silverstein

THROUGHOUT THE 1990s, ARCH ACCEPTED a number of exciting projects that tested its capabilities and challenged the creativity of its employees. In some cases, the job presented a challenge based solely upon sheer product volume. In other cases, unique design requests tested the engineering skills and the meticulous craftsmanship of talented Arch employees who transformed the project from paper to physical reality.

In almost every case, the prospective client would call Arch and say they were referred to Arch because of its innovative reputation and ability to conquer complex and unusual jobs.

The examples that follow illustrate a sampling of Arch's most innovative and extraordinary projects to date. Regardless of the size or complexity of the special project, Arch's regular customers always received their orders as promised because of Arch's strong commitment to meeting the needs of its valued and loyal customers.

Starrett City

In the early 1990s, Leon and Bobby Silverstein were asked if they could supply thermal balcony doors for an apartment complex under construction in Starrett City, New York. At the time, patio sliding doors were the fashion but offered little protection from the cold and heat. The architect on this job wanted a two-tone French-style thermal door instead. "No one was making a door that way," Leon said, but Arch had the capacity to design such a door. Leon told his contact he would consider taking on the job if the order proved substantial enough.[1]

It turned out that the architectural firm needed 1,000 thermal doors, so Arch agreed to the job. "That was one of the biggest orders we have had," Leon explained.[2] In total, Arch would need to deliver eight truckloads of doors to New York over a six-month period. Because Arch's truck drivers, like all of its employees, are nonunion, they experienced trouble on their way to Starrett City, when union teamsters on the jobsite forced the truck to a halt before a bridge crossing. Leon and Bobby were alerted and contacted one of their suppliers in New York who, according to Leon, "knew the right people."[3]

Leon and Bobby learned that if they paid $400 in cash per truck, the men would permit the trucks to cross the bridge. Leon commented, "We felt dirty.

Opposite: After successfully designing the custom thermal balcony doors for the apartment complex in Starrett City, New York, Arch created this thermal door as a standard item.

Above: Arch fabricated the oversized interior glasswork for the Time Warner Building, one of the most prestigious projects the company has tackled in its history. *(© Image from BigStockPhoto.com.)*

We did sell other thermal doors in New York, but it was on our terms. Over time, we stopped doing business there."[4]

Having to pay off people to get the product into New York set the tone for the rest of the project. Leon had arranged for payment from the architect within 30 days of each delivery, but payments did not arrive in a timely fashion. As soon as the last shipment of doors left the Arch warehouse, the architect began making vague complaints about the quality of the doors. Leon worried that future payments might not be forthcoming.

To protect Arch, Leon coordinated the removal of 14 doors from the final truckload that would be held by Arch in lieu of payment for the balance of the order. The confiscated doors, as chance would have it, were intended for the penthouse. The final truckload, minus those 14 doors, was delivered. Apparently, no one at the construction site noticed that the doors were missing.

Thirty days later, Leon called for his money. Two weeks passed before Leon received a response from the architect. According to Leon, "The architect said, 'We are 14 doors short and your manifest shows you shipped them all.' I said, 'I stopped the truck in New Jersey and removed the doors since I had a feeling you were not going to pay. If you want the 14 doors, you have to pay the account in full.'"[5]

The architect wanted the penthouse doors and paid the account in full. Leon, in turn, delivered the doors as promised.

Time Warner Building

Since the mid-1990s, Arch has had the capacity to fabricate oversized glass. In 2002, Glass Alum, a Miami-based contract glazier, approached Arch with a job to fabricate 100-by-100–inch blocks of oversized glass for one of its customers, Time Warner. "Glass Alum was one of the few companies in the country that could do monumental work and do it well. It actually handles a lot of work in New York. By fabricating for Glass Alum, we wouldn't have to deal with New York directly," Leon explained.[1]

Only three furnaces in the country could accommodate such a large piece of glass, and Arch Aluminum & Glass owned two of those furnaces. Because of its size, this job would be very difficult to complete successfully. Leon said:

> *We had to do it well, since Time Warner—the client—was expecting quality work. We were leery of the project because Time Warner was not a regular customer of ours. We decided to make a mock piece of glass as a sample for their Time Warner Building. We told them, if you like the sample, we will accept your order. If you don't, no problem, we will throw the sample away and no harm done.*[2]

Time Warner approved the sample and placed an order consisting of hundreds of pieces of glass.

Donny Anderson, the manager of the Tamarac facility, took production of this order upon himself. According to Ricky Silverstein, "Donny took himself out of his usual job to run $300,000 worth of glass over two to three weeks. For four hours every day, that was his job. He went from plant manager to oven operator."[3]

Although fabricated by machine, the operator makes the difference in the quality of the tempered glass. "The process has to be watched carefully because what you think will happen to the glass may not happen," Leon explained. "If you made the product and you had zero rejects, it was a very good order. If you made the product and every second piece you made had to be thrown away, it wasn't a very good order."[4]

Anderson provided strict quality control. "When we had other people do it, the quality wasn't any good. When Donny did it, it was like magic, it was always good," Leon said. "We shipped four or five truckloads of glass to Glass Alum, and we did not have one remake or one complaint. We were paid timely. It was a great job. It was prestigious. We provided the glass for the Time Warner Building in New York City. And it was good to know we could work in New York through a Miami firm and not have to deal with gangsters."[5]

Farhat Aluminum

Harold Dudoff, Sydney and Bobby Silverstein's longtime friend who had initiated overseas work for them back in the days of Architectural Glass in Miami, continued to negotiate deals abroad well into the 1990s. Dudoff, while in the Middle East, made arrangements for Arch to supply glass and aluminum for Farhat Aluminum company, which served as the general contractors for a mixed-use construction complex in Abu Dhabi, United Arab Emirates.

The deal was contingent on Dudoff's involvement, so he remained in the Middle East to maintain communications with the contractors at Farhat

Left: Donny Anderson, plant manager for the Tamarac facility, stood watch over the tempering of glass for the Time Warner project. He provided excellent quality control to help make the job a highly successful one. *(Photo courtesy of IDS Photography.)*

Below: Arnie Studenski traveled to the United Arab Emirates to serve as the intermediary for Arch and the contractors at Farhat Aluminum Company after the untimely death of Harold Dudoff.

Arch provided all the glass exterior windows and doors for the impressive-looking Atlantis Hotel on Paradise Island in the Bahamas. *(Photo courtesy of PPG Industries, Inc.)*

Aluminum Company and the project managers at Arch. "The construction complex required an elaborate design. A lot of the glass came to various points that created the illusion of turbans," Leon explained.[1]

After production on the job began, Dudoff became ill and died unexpectedly. Although saddened by Dudoff's death, Leon and Bobby were determined to finish the job he had arranged. Rather than complicate matters by acknowledging Dudoff's death, Leon and Bobby sent Arnie Studenski to assist with communications and made excuses when the contractors asked for Dudoff. They simply would say that Dudoff was unavailable. The contractors ultimately stopped asking for Dudoff and allowed Studenski to handle any problems that arose.

After completing the job, Leon communicated the news of Dudoff's death to the contractors. "Unfortunately, we didn't receive additional work from that area because it is not what you do, or how well you do it, but it is who you know," Leon said.[2]

The Atlantis Hotel

Construction began on the Atlantis Hotel on Paradise Island in the mid-1990s. At the time, John Trico, of Commonwealth Glass, a longtime customer of Arch and a Bahamian contractor, contacted Leon to request an estimate from Arch to provide all the glass for what he called the skin of the building—meaning all the exterior doors and windows of the hotel.

Arch provided the quote to Trico, and he received the contract. The job itself was not complex, but it was massive. According to Leon, "It was a signature job because of the sheer volume of project size, and the visibility of the building was great."[1]

Anheuser-Busch Brewing Company

St. Louis is known as the Gateway to the West, and it is known for its famous brewery. Beer commercials have touted the city for housing the world's largest brewery—Anheuser-Busch. The facility is registered as a national historic landmark, boasting 137 structures, most of which were built before the turn of the 20th century.

When renovations began on one of the buildings, the architect wanted to replace the original entranceway of massive, finely milled wooden doors, which had rotted away over the years, with aluminum replicas. Leon commented that Arch took the job because of the challenge it presented. "It was fun to do because it tested the limits of what could and could not be done in aluminum. We actually designed everything to duplicate the doors exactly," he said.[1]

The challenges began with regard to the large size and large number of extrusion shapes necessary. More than a dozen specialized extrusion shapes were required, with some of those pieces as large as 12 inches in width. Almost every item required a custom design. Leon said:

> *It really tested the craftsmanship of our technicians. They were literally playing with aluminum like a mill workman would play with wood. Aluminum doesn't allow itself to be played with like that. Every piece of aluminum was hand-cut, hand-milled, and hand-mitered for each opening.*
>
> *Everything took five times longer than we anticipated. It is one of those jobs we didn't expect to make any money on. But we were very proud of the job when we were done.*[2]

Universal Studios

Universal Studios in Orlando, Florida, contracted with Arch to build a large seamless piece of glass for use in its *Ghostbusters* attraction. The glass would refract and reflect lasers, giving the illusion of the Ghostbusters fighting a ghost.

To test the type of glass needed, Arch created a small prototype. Attraction creators successfully tested the sample glass and proceeded to place an order for the actual glass. The order included 11 units of 130-by-204–inch glass, the largest size of laminated glass available. It also required 11 units of 96-by-130–inch glass.

The specialized glass needed to be made from liquid laminate. Leon explained:

> *We knew how to handle glass. We did not know how to handle liquid laminate that big. We had to spend $7,500 on a special table we needed for the liquid laminate to be able to flow.*[1]

By successfully creating the glass, Arch once again proved to be the innovative leader by producing a product no one else could handle.

Entourage Martini Bar

An Arch customer and glazier in Chicago was awarded a unique project that involved constructing the façade for a new martini bar to be called Entourage. Keeping with the martini theme, the architect wanted a gigantic three-dimensional martini shaker built into the building's exterior. The glazier provided Jeff Ziesche, director of Arch Aluminum, with a sketch of the idea and asked him if Arch could design and fabricate the aluminum extrusions for the job.

Starting in July 2005, Ziesche's crew began creating the conceptual drawings for the dies and extrusions. Once completed, the glazier, architect, and owner reviewed and approved the plan. In the case of this project, Arch needed to design and create dies and extrusions in unusual shapes and lengths to accommodate the martini shaker plan. Then, the extrusion technicians employed the traditional extrusion process, which involved heating the aluminum to a malleable temperature so it could be squeezed through the custom-made metal dies that fabricated the parts in uniform cross sections.

Arch completed the job and delivered the goods for this custom project in record time. Ziesche proudly recalled, "We had materials to the glazier by December 2005."[1]

The martini shaker finished product bears a strong resemblance to the original drawing. *(Photo courtesy of IDS Photography.)*

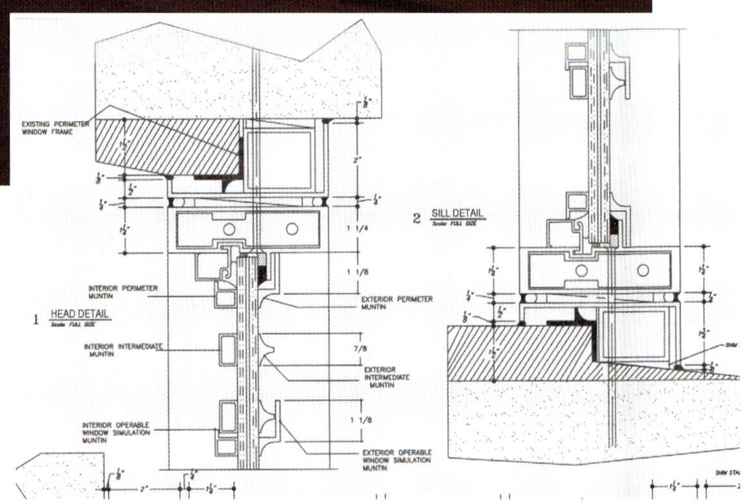

Above: The Gato building as it appears after renovations were complete. *(Photo courtesy of IDS Photography.)*

Right: This blueprint illustrates the various details associated with the windows and frames for the Gato building project.

The Gato Cigar Factory

One of Arch's most unusual special projects involved the renovation of the Gato Cigar factory, built by Cuban immigrant Eduardo Gato in Key West, Florida. Several years ago, the local government took possession of the building and decided to renovate it. As a result, the city was required to meet the revised hurricane codes as well as the historic restoration regulations.

The city worked with Wind-Tech Products, Inc., a window distributor and installer, to handle the renovations. Wind-Tech, in turn, sought out a firm that could handle both historical and environmental challenges and awarded Arch Aluminum & Glass the project in 2002. Jeff Ziesche spearheaded the project.

Located in a hurricane-prone area, the building had experienced both adverse environmental conditions as well as years of neglect. The new façade would need to tolerate hurricane winds and gales, high humidity, and unrelenting high temperatures. Also, the new aluminum window frames needed to replicate the original hunter green wooden frames that once graced the building's front.

Wind-Tech built a duplicate of the original windows out of wood, including the original framing system. Arch then matched the delicately designed wooden details with its custom-designed aluminum frames and muntins (framing parts of the window that divide the panes of glass). Arch developed a custom powder-coat finish that matched the original hunter green color exactly.[1]

To protect the building from hurricanes, Arch used its Impact 3000 glass, which exceeds current hurricane code requirements. As the project neared completion, the state historic preservation field officer flew down from Tallahassee, Florida, to verify and approve the authentic appearance of the historic building, including its new aluminum frames and windows. The renovated building was found to be both functional and faithful to the original design.[2]

A Timeline on the Evolution

1978 Arch opens in Miami, FL

1981 Arch moves to Ft. Lauderdale, FL

1983 Sarasota, FL opens

1985 Ft. Myers, FL opens

1986 Headquarters move to Pompano Beach, FL

1987 First tempering plant opens in Orlando, FL

1988 2nd tempering plant opens in Austell, GA

1992 Laminating plant opens in Ft. Pierce, FL

1994 Indianapolis, IN opens

Arch purchases Amarlite in Atlanta, GA

Arch Indianapolis supplied the interior for the First Federal Savings Bank in Louisville, Kentucky.

1978

Arch Kansas provided the glass for this Gallup University building in Omaha, Nebraska.

1995 Kansas City, KS opens

Arch Mirror opens in Bettendorf, IA

Arch Mirror opens in Ft. Pierce, FL

1996 Norcross, GA opens

Villa Rica, GA opens

1997 Youngsville, NC opens

1999 Jackson, MS opens

Arch Jackson supplied the glass and aluminum for the Performance Contractors building in Baton Rouge, Louisiana.

Arch Phoenix opens in 2000.

2000 Bridgeport, NJ opens

Waukesha, WI opens (with 2nd laminated facility)

Phoenix, AZ opens

# of Arch Aluminum & Glass

PHOTO COURTESY OF IDS PHOTOGRAPHY.

In 2004, Arch Phoenix supplied the glass for this Arizona State University building.

2002 Arch acquires American Glassmith in Columbus, OH

2004 Denver, CO opens

Arch acquires Glass-Tex in Houston, TX

PHOTO COURTESY OF IDS PHOTOGRAPHY.

Arch Phoenix supplied the materials for the Westin in Scottsdale, Arizona.

2001 Akron, OH opens

Arch Mirror West opens in California

Dallas, TX opens

2003 Arch Extrusions opens in Miami, FL

Arch acquires Trident in City of Industry, CA (3rd laminating facility)

Arch opens 4th laminating facility in Tamarac, FL

Arch Deco in Columbus, Ohio, supplied the blue decorative glass for the Virgin Airline Club at the San Francisco Airport.

This FedEx building was constructed with materials from Arch Akron.

2008

2005 Arch acquires Arch Aluminum of Texas

Arch moves Columbus facility to new building and adds 5th laminating facility

2006 Arch moves Dallas, TX operations to Ft. Worth and adds 7th laminating facility

Arch adds 8th laminating facility in Villa Rica, GA

Arch opens in Las Vegas, NV

Arch adds 2nd extrusion line in Miami, FL

Shortly after its expansion in 2006, Arch Orlando supplied the material for this new Orange County Sheriff facility.

Arch Deco opens a new facility with laminating capacity in Columbus, Ohio, in 2005.

2007 Arch opens in Rogers, MN

Arch exceeds $350M in sales

2006 Arch opens in Salt Lake City, UT

Arch adds 6th laminating facility in Orlando, FL

Arch Salt Lake City during construction.

2008 Arch acquires a facility in Tulsa, OK

NOTES TO SOURCES

Chapter One

1. *The Chronicle of the 20th Century* (Mount Kisco, New York: Chronicle Publications, 1987), page 403.
2. Dan Klein and Ward Lloyd, *The History of Glass* (London: Orbin Publishing, Limited, 1984), pages 222–224.
3. James H. Wasley and Michael Utzinger, "Vital Signs Curriculum Materials Project" (Berkeley, California: College of Environmental Design, University of California, Berkeley, 1996), Part 2, 24, http://www.arch.ced.berkeley.edu/vitalsigns/res/index.html (accessed 4 January 2006).
4. James H. Wasley and Michael Utzinger, "Vital Signs Curriculum Materials Project" (Berkeley, California: College of Environmental Design, University of California, Berkeley, 1996), Part 2, 24, http://www.arch.ced.berkeley.edu/vitalsigns/res/index.html (accessed 4 January 2006).
5. *Preservation Brief 12: The Preservation of Historic Pigmented Structural Glass* (Technical Preservation Services, National Park Service, U.S. Department of the Interior), http://cr.nps.gov/hps/tps/briefs/brief12.htm (accessed 4 January 2006).
6. "The Age of Aluminum: A Man, a Metal, a Vision ... and the Rest Is History," Aluminum Now Online, http://www.aluminum.org/ANTemplate.cfm (accessed 2 January 2006).
7. Bobby Silverstein, interview by Ann Gossy and Linda A. Firestone, digital recording, 12 January 2006, Write Stuff Enterprises, Inc.
8. Beani Silverstein, interview by Ann Gossy and Linda A. Firestone, digital recording, 17 January 2006, Write Stuff Enterprises, Inc.
9. *1898–1998*, Perilstein Centennial Pamphlet (Philadelphia, Perilstein Distributing Group, 1998), page 5.
10. Max Perilstein, e-mail message to Linda A. Firestone, 12 January 2006.
11. Bobby Silverstein, interview by Ann Gossy and Linda A. Firestone, digital recording, 12 January 2006, Write Stuff Enterprises, Inc.
12. Bobby Silverstein, interview by Ann Gossy and Linda A. Firestone,

digital recording, 12 January 2006, Write Stuff Enterprises, Inc.

13. Paul and Ruth Reifman, interview by Jeffrey L. Rodengen, digital recording, 20 April 2006, Write Stuff Enterprises, Inc.
14. Paul and Ruth Reifman, interview by Jeffrey L. Rodengen, digital recording, 20 April 2006, Write Stuff Enterprises, Inc.
15. Bobby Silverstein, interview by Ann Gossy and Linda A. Firestone, digital recording, 12 January 2006, Write Stuff Enterprises, Inc.
16. Bobby Silverstein, interview by Ann Gossy and Linda A. Firestone, digital recording, 12 January 2006, Write Stuff Enterprises, Inc.
17. Bobby Silverstein, interview by Ann Gossy and Linda A. Firestone, digital recording, 12 January 2006, Write Stuff Enterprises, Inc.
18. Bobby Silverstein, interview by Ann Gossy and Linda A. Firestone, digital recording, 10 January 2006, Write Stuff Enterprises, Inc.
19. Beani Silverstein, interview by Ann Gossy and Linda A. Firestone, digital recording, 17 January 2006, Write Stuff Enterprises, Inc.
20. Beani Silverstein, interview by Jeffrey L. Rodengen, digital recording, 18 October 2005, Write Stuff Enterprises, Inc.
21. E-mail correspondence with Arch Aluminum & Glass, 2 August 2006.
22. Beani Silverstein, interview by Jeffrey L. Rodengen, digital recording, 18 October 2005, Write Stuff Enterprises, Inc.
23. Beani Silverstein, interview by Jeffrey L. Rodengen, digital recording, 18 October 2005, Write Stuff Enterprises, Inc.
24. Beani Silverstein, interview by Jeffrey L. Rodengen, digital recording, 18 October 2005, Write Stuff Enterprises, Inc.

Chapter Two

1. Paul Parcell, interview by Jeffrey L. Rodengen, digital recording, 18 October 2005, Write Stuff Enterprises, Inc.
2. Arnie Studenski, interview by Ann Gossy and Linda A. Firestone, digital recording, 6 February 2006, Write Stuff Enterprises, Inc.
3. Roger Collette, interview by Jeffrey L. Rodengen, digital recording, 19 October 2006, Write Stuff Enterprises, Inc.
4. Bobby Silverstein, remarks made to Linda A. Firestone, handwritten comments, 9 February 2006, Write Stuff Enterprises, Inc.
5. Bobby Silverstein, remarks made to Linda A. Firestone, handwritten comments, 9 February 2006, Write Stuff Enterprises, Inc.
6. Bobby Silverstein, interview by Ann Gossy and Linda A. Firestone, digital recording, 12 January 2006, Write Stuff Enterprises, Inc.
7. Bobby Silverstein, interview by Jeffrey L. Rodengen, digital recording, 18 October 2005, Write Stuff Enterprises, Inc.
8. Bobby Silverstein, interview by Jeffrey L. Rodengen, digital recording, 18 October 2005, Write Stuff Enterprises, Inc.
9. Bobby Silverstein, interview by Ann Gossy and Linda A. Firestone, digital recording, 12 January 2006, Write Stuff Enterprises, Inc.
10. Beani Silverstein, interview by Ann Gossy and Linda A. Firestone,

digital recording, 17 January 2006, Write Stuff Enterprises, Inc.
11. Beani Silverstein, interview by Ann Gossy and Linda A. Firestone, digital recording, 17 January 2006, Write Stuff Enterprises, Inc.
12. Beani Silverstein, interview by Ann Gossy and Linda A. Firestone, digital recording, 17 January 2006, Write Stuff Enterprises, Inc.
13. Bobby Silverstein, interview by Ann Gossy and Linda A. Firestone, digital recording, 10 January 2006, Write Stuff Enterprises, Inc.
14. Bobby Silverstein, interview by Ann Gossy and Linda A. Firestone, digital recording, 10 January 2006, Write Stuff Enterprises, Inc.
15. Bobby Silverstein, interview by Jeffrey L. Rodengen, digital recording, 18 October 2005, Write Stuff Enterprises, Inc.
16. Beani Silverstein, interview by Ann Gossy and Linda A. Firestone, digital recording, 17 January 2006, Write Stuff Enterprises, Inc.
17. Leon Silverstein, interview by Ann Gossy and Linda A. Firestone, digital recording, 9 January 2006, Write Stuff Enterprises, Inc.
18. Bobby Silverstein, interview by Ann Gossy and Linda A. Firestone, digital recording, 10 January 2006, Write Stuff Enterprises, Inc.
19. Bobby Silverstein, interview by Ann Gossy and Linda A. Firestone, digital recording, 10 January 2006, Write Stuff Enterprises, Inc.
20. Beani Silverstein, interview by Ann Gossy and Linda A. Firestone, digital recording, 17 January 2006, Write Stuff Enterprises, Inc.
21. Bobby Silverstein, interview by Linda A. Firestone, handwritten notes, 22 February 2006, Write Stuff Enterprises, Inc.

Chapter Two Sidebar: Road Less Traveled

1. Roger Collette, interview by Jeffrey L. Rodengen, digital recording, 19 October 2005, Write Stuff Enterprises, Inc.

Chapter Two Sidebar: A Testament to Quality Service

1. Handwritten list created by Roger Collette, reviewed by Bobby Silverstein, received by Linda A. Firestone, 28 February 2006.

Chapter Two Sidebar: Memory Lane

1. Leon Silverstein, interview by Ann Gossy and Linda A. Firestone, digital recording, 14 February 2006, Write Stuff Enterprises, Inc.

Chapter Three

1. Bobby Silverstein, interview by Jeffrey L. Rodengen, digital recording, 18 October 2005, Write Stuff Enterprises, Inc.
2. Bobby Silverstein, interview by Jeffrey L. Rodengen, digital recording, 18 October 2005, Write Stuff Enterprises, Inc.
3. Bobby Silverstein, interview by Ann Gossy, digital recording, 19 December 2005, Write Stuff Enterprises, Inc.
4. Bobby Silverstein, interview by Jeffrey L. Rodengen, digital recording, 18 October 2005, Write Stuff Enterprises, Inc.
5. Bobby Silverstein, interview by Jeffrey L. Rodengen, digital recording, 18 October

2005, Write Stuff Enterprises, Inc.
6. Beani Silverstein, interview by Ann Gossy and Linda A. Firestone, digital recording, 17 January 2006, Write Stuff Enterprises, Inc.
7. Beani Silverstein, interview by Ann Gossy and Linda A. Firestone, digital recording, 17 January 2006, Write Stuff Enterprises, Inc.
8. Bobby Silverstein, interview by Ann Gossy, digital recording, 19 December 2005, Write Stuff Enterprises, Inc.
9. Mike Rogers, interview by Ann Gossy and Linda A. Firestone, digital recording, 25 January 2006, Write Stuff Enterprises, Inc.
10. Mike Rogers, interview by Ann Gossy and Linda A. Firestone, digital recording, 25 January 2006, Write Stuff Enterprises, Inc.
11. Don Beaver, interview by Ann Gossy and Linda A. Firestone, digital recording, 10 March 2006, Write Stuff Enterprises, Inc.
12. Don Beaver, interview by Ann Gossy and Linda A. Firestone, digital recording, 10 March 2006, Write Stuff Enterprises, Inc.
13. Bobby Silverstein, interview by Ann Gossy and Linda A. Firestone, digital recording, 12 January 2006, Write Stuff Enterprises, Inc.
14. Ray London, interview by Ann Gossy and Linda A. Firestone, digital recording, 16 March 2006, Write Stuff Enterprises, Inc.
15. Bobby Silverstein, interview by Ann Gossy and Linda A. Firestone, digital recording, 14 February 2006, Write Stuff Enterprises, Inc.
16. Bobby Silverstein, interview by Ann Gossy and Linda A. Firestone, digital recording, 14 February 2006, Write Stuff Enterprises, Inc.
17. Beani Silverstein, interview by Ann Gossy and Linda A. Firestone, digital recording, 17 January 2006, Write Stuff Enterprises, Inc.
18. Beani Silverstein, interview by Ann Gossy and Linda A. Firestone, digital recording, 17 January 2006, Write Stuff Enterprises, Inc.
19. Beani Silverstein, interview by Ann Gossy and Linda A. Firestone, digital recording, 17 January 2006, Write Stuff Enterprises, Inc.
20. Arnold Kramer, interview by Jeffrey L. Rodengen, digital recording, 18 May 2006, Write Stuff Enterprises, Inc.
21. Bobby Silverstein, interview by Jeffrey L. Rodengen, digital recording, 18 October 2005, Write Stuff Enterprises, Inc.

Chapter Four

1. Bobby Silverstein, interview by Jeffrey L. Rodengen, digital recording, 18 October 2005, Write Stuff Enterprises, Inc.
2. Bobby Silverstein, interview by Jeffrey L. Rodengen, digital recording, 18 October 2005, Write Stuff Enterprises, Inc.
3. Bobby Silverstein, interview by Jeffrey L. Rodengen, digital recording, 18 October 2005, Write Stuff Enterprises, Inc.
4. Bobby Silverstein, interview by Jeffrey L. Rodengen, digital recording, 18 October 2005, Write Stuff Enterprises, Inc.
5. Alan Moses, interview by Ann Gossy and Linda A. Firestone, digital recording, 7 February

2006, Write Stuff Enterprises, Inc.
6. Douglas Cove, interview by Jeffery L. Rodengen, digital recording, 1 December 2005, Write Stuff Enterprises, Inc.
7. Bobby Silverstein, interview by Jeffrey L. Rodengen, digital recording, 18 October 2005, Write Stuff Enterprises, Inc.
8. Bobby Silverstein, interview by Jeffrey L. Rodengen, digital recording, 18 October 2005, Write Stuff Enterprises, Inc.
9. Bobby Silverstein, interview by Jeffrey L. Rodengen, digital recording, 18 October 2005, Write Stuff Enterprises, Inc.
10. Bobby Silverstein, interview by Jeffrey L. Rodengen, digital recording, 18 October 2005, Write Stuff Enterprises, Inc.
11. Bobby Silverstein, interview by Jeffrey L. Rodengen, digital recording, 18 October 2005, Write Stuff Enterprises, Inc.
12. Douglas Cove, interview by Jeffery L. Rodengen, digital recording, 1 December 2005, Write Stuff Enterprises, Inc.
13. Leon Silverstein, interview by Ann Gossy and Linda A. Firestone, digital recording, 17 January 2006, Write Stuff Enterprises, Inc.
14. Leon Silverstein, interview by Ann Gossy and Linda A. Firestone, digital recording, 17 January 2006, Write Stuff Enterprises, Inc.
15. Richard Davis, interview by Jeffrey L. Rodengen, digital recording, 19 October 2005, Write Stuff Enterprises, Inc.
16. Don Beaver, interview by Ann Gossy and Linda A. Firestone, digital recording, 10 March 2006, Write Stuff Enterprises, Inc.
17. Leon Silverstein, interview by Ann Gossy and Linda A. Firestone, digital recording, 17 January 2006, Write Stuff Enterprises, Inc.
18. Leon Silverstein, interview by Ann Gossy and Linda A. Firestone, digital recording, 14 February 2006, Write Stuff Enterprises, Inc.
19. Kevin Aubuchon, interview by Jeffrey L. Rodengen, digital recording, 18 October 2005, Write Stuff Enterprises, Inc.
20. Donna Aubuchon, interview by Ann Gossy and Linda A. Firestone, digital recording, 31 January 2006, Write Stuff Enterprises, Inc.
21. Kevin Aubuchon, interview by Ann Gossy and Linda A. Firestone, digital recording, 7 April 2006, Write Stuff Enterprises, Inc.
22. Kevin Aubuchon, interview by Jeffrey L. Rodengen, digital recording, 18 October 2005, Write Stuff Enterprises, Inc.
23. Kevin Aubuchon, interview by Jeffrey L. Rodengen, digital recording, 18 October 2005, Write Stuff Enterprises, Inc.
24. Kevin Aubuchon, interview by Jeffrey L. Rodengen, digital recording, 18 October 2005, Write Stuff Enterprises, Inc.
25. Kevin Aubuchon, interview by Jeffrey L. Rodengen, digital recording, 18 October 2005, Write Stuff Enterprises, Inc.
26. "Arch Aluminum Advances," *Float Lines*, a publication of Ford's Glass Division, No. 3, Fall 1986.

Chapter Four Sidebar: How Arch Got Its Name

1. Beani Silverstein, interview by Ann Gossy and Linda

A. Firestone, digital recording, 17 January 2006, Write Stuff Enterprises, Inc.

Chapter Four Sidebar: Keeping Promises

1. Arnie Studenski, interview by Ann Gossy and Linda A. Firestone, digital recording, 2 February 2006, Write Stuff Enterprises, Inc.

Chapter Four Sidebar: Rescue Mission

1. Roger Collette, interview by Jeffrey L. Rodengen, digital recording, 19 October 2005, Write Stuff Enterprises, Inc.

Chapter Four Sidebar: Building Loyalty

1. Ray London, interview by Jeffrey L. Rodengen, digital recording, 2 December 2005, Write Stuff Enterprises, Inc.

Chapter Five

1. "Glass: A Clear Vision for a Bright Future," U.S. Department of Energy, January 1996, http://www.eere.energy.gov/industry/glass/pdfs/glass_vision.pdf (accessed February 2006).
2. Gerald Ashley, interview by Jeffrey L. Rodengen, digital recording, 6 April 2005, Write Stuff Enterprises, Inc.
3. Gabe Garcia, interview by Jeffrey L. Rodengen, digital recording, 18 October 2005, Write Stuff Enterprises, Inc.
4. Neil Murphy, interview by Jeffrey L. Rodengen, digital recording, 2 December 2005, Write Stuff Enterprises, Inc.
5. Art Jackson, interview by Ann Gossy and Linda A. Firestone, digital recording, 19 April 2006, Write Stuff Enterprises, Inc.
6. Leon Silverstein, interview by Linda A. Firestone, handwritten notes, 19 April 2006, Write Stuff Enterprises, Inc.
7. Leon Silverstein, interview by Linda A. Firestone, handwritten notes, 19 April 2006, Write Stuff Enterprises, Inc.
8. Leon Silverstein, interview by Linda A. Firestone, handwritten notes, 19 April 2006, Write Stuff Enterprises, Inc.
9. Leon Silverstein, interview by Linda A. Firestone, handwritten notes, 19 April 2006, Write Stuff Enterprises, Inc.
10. Paul Parcell, interview by Jeffrey L. Rodengen, 18 October 2005, digital recording, Write Stuff Enterprises, Inc.
11. Paul Parcell, interview by Jeffrey L. Rodengen, 18 October 2005, digital recording, Write Stuff Enterprises, Inc.
12. Paul Parcell, interview by Jeffrey L. Rodengen, 18 October 2005, digital recording, Write Stuff Enterprises, Inc.
13. Leon Silverstein, interview by Linda A. Firestone, digital recording, 27 April 2006, Write Stuff Enterprises, Inc.
14. Kevin Aubuchon, interview Jeffrey L. Rodengen, digital recording, 18 October 2006, Write Stuff Enterprises, Inc.
15. Garth Goodrich, interview by Jeffrey L. Rodengen, digital recording, 5 April 2006, Write Stuff Enterprises, Inc.
16. Leon Silverstein, interview by Ann Gossy and Linda A. Firestone, digital recording, 17 January 2006, Write Stuff Enterprises, Inc.
17. Mary and Numa Handi, interview by Linda A. Firestone, digital recording, 24 April 2006, Write Stuff Enterprises, Inc.
18. Leon Silverstein, interview by Linda A. Firestone,

digital recording, 27 April 2006, Write Stuff Enterprises, Inc.
19. Leon Silverstein, interview by Linda A. Firestone, digital recording, 27 April 2006, Write Stuff Enterprises, Inc.
20. "Hurricane Building Codes," Miami Dade Department of Building Codes, http://www.miamidade.gov/buildingcode/au_product_control.asp (accessed 4 June 2006).
21. Leon Silverstein, interview by Linda A. Firestone, digital recording, 27 April 2006, Write Stuff Enterprises, Inc.
22. "Hurricane Building Codes," Miami Dade Department of Building Codes, http://www.miamidade.gov/buildingcode/au_product_control.asp (accessed 4 June 2006).
23. Leon Silverstein, interview by Ann Gossy and Linda A. Firestone, digital recording, 17 January 2006, Write Stuff Enterprises, Inc.
24. Leon Silverstein, interview by Ann Gossy and Linda A. Firestone, digital recording, 17 January 2006, Write Stuff Enterprises, Inc.
25. Leon Silverstein, interview by Ann Gossy and Linda A. Firestone, digital recording, 17 January 2006, Write Stuff Enterprises, Inc.
26. Leon Silverstein, interview by Ann Gossy and Linda A. Firestone, digital recording, 17 January 2006, Write Stuff Enterprises, Inc.
27. Leon Silverstein, interview by Ann Gossy and Linda A. Firestone, digital recording, 17 January 2006, Write Stuff Enterprises, Inc.
28. Leon Silverstein, interview by Ann Gossy and Linda A. Firestone, digital recording, 17 January 2006, Write Stuff Enterprises, Inc.
29. Leon Silverstein, interview by Linda A. Firestone, digital recording, 27 April 2006, Write Stuff Enterprises, Inc.
30. Leon Silverstein, interview by Ann Gossy and Linda A. Firestone, digital recording, 17 January 2006, Write Stuff Enterprises, Inc.
31. Leon Silverstein, interview by Ann Gossy and Linda A. Firestone, digital recording, 17 January 2006, Write Stuff Enterprises, Inc.
32. Leon Silverstein, interview by Ann Gossy and Linda A. Firestone, digital recording, 17 January 2006, Write Stuff Enterprises, Inc.
33. Charles Cumston, "What Does Leon Think," *Glass Magazine*, March 2000, page 52.
34. Leon Silverstein, interview by Ann Gossy and Linda A. Firestone, digital recording, 17 January 2006, Write Stuff Enterprises, Inc.
35. Leon Silverstein, interview by Linda A. Firestone, digital recording, 27 April 2006, Write Stuff Enterprises, Inc.
36. Leon Silverstein, interview by Ann Gossy and Linda A. Firestone, digital recording, 17 January 2006, Write Stuff Enterprises, Inc.
37. Leon Silverstein, interview by Ann Gossy and Linda A. Firestone, digital recording, 17 January 2006, Write Stuff Enterprises, Inc.
38. Leon Silverstein, interview by Ann Gossy and Linda A. Firestone, digital recording, 17 January 2006, Write Stuff Enterprises, Inc.
39. Leon Silverstein, interview by Ann Gossy and Linda A. Firestone, digital recording, 17 January 2006, Write Stuff Enterprises, Inc.

40. Leon Silverstein, interview by Ann Gossy and Linda A. Firestone, digital recording, 17 January 2006, Write Stuff Enterprises, Inc.

Chapter Five Sidebar: The Growth of the Green Glass Industry

1. "Easy Being Green?" *U.S. Glass* magazine, February 2005.

Chapter Five Sidebar: What's in a Name?

1. Paula Coben, interview by Linda A. Firestone, digital recording, 17 April 2006, Write Stuff Enterprises, Inc.

Chapter Five Sidebar: Hurricane Prompts Improved Standards

1. Joseph Minor, "Lessons Learned from Failures of the Building Envelope in Windstorms," *Journal of Architectural Engineering*, March 2005.

Chapter Five Sidebar: Success at Auction

1. Leon Silverstein, interview by Linda A. Firestone, digital recording, 17 January 2006, Write Stuff Enterprises, Inc.

Chapter Five Sidebar: Changing with the Times

1. Ron Brock, interview by Ann Gossy and Linda A. Firestone, digital recording, 4 April 2006, Write Stuff Enterprises, Inc.

Chapter Five Sidebar: Jumbo Glass

1. Leon Silverstein, interview by Linda A. Firestone, digital recording, 27 April 2006, Write Stuff Enterprises, Inc.

Chapter Five Sidebar: Weathering the Storm

1. Mike Rogers, interview by Linda Firestone, digital recording, 25 January 2006, Write Stuff Enterprises, Inc.

Chapter Six

1. Leon Silverstein, interview by Linda A. Firestone, digital recording, 4 May 2006, Write Stuff Enterprises, Inc.
2. Leon Silverstein, interview by Linda A. Firestone, digital recording, 4 May 2006, Write Stuff Enterprises, Inc.
3. Leon Silverstein, interview by Linda A. Firestone, digital recording, 4 May 2006, Write Stuff Enterprises, Inc.
4. Leon Silverstein, interview by Linda A. Firestone, digital recording, 4 May 2006, Write Stuff Enterprises, Inc.
5. Leon Silverstein, interview by Linda A. Firestone, digital recording, 4 May 2006, Write Stuff Enterprises, Inc.
6. Leon Silverstein, interview by Linda A. Firestone, digital recording, 4 May 2006, Write Stuff Enterprises, Inc.
7. Leon Silverstein, interview by Linda A. Firestone, digital recording, 27 April 2006, Write Stuff Enterprises, Inc.
8. Leon Silverstein, interview by Linda A. Firestone, digital recording, 27 April 2006, Write Stuff Enterprises, Inc.
9. Leon Silverstein, interview by Linda A. Firestone, digital recording, 27 April 2006, Write Stuff Enterprises, Inc.
10. Leon Silverstein, interview by Linda A. Firestone, digital recording, 27 April 2006, Write Stuff Enterprises, Inc.
11. Leon Silverstein, interview by Linda A. Firestone, digital recording, 27 April 2006, Write Stuff Enterprises, Inc.

12. Leon Silverstein, interview by Linda A. Firestone, digital recording, 27 April 2006, Write Stuff Enterprises, Inc.
13. Leon Silverstein, interview by Linda A. Firestone, digital recording, 17 January 2006, Write Stuff Enterprises, Inc.
14. Leon Silverstein, interview by Linda A. Firestone, digital recording, 17 January 2006, Write Stuff Enterprises, Inc.
15. Leon Silverstein, interview by Linda A. Firestone, digital recording, 17 January 2006, Write Stuff Enterprises, Inc.
16. Don Powers, interview by Jeffrey L. Rodengen, digital recording, 6 April 2006, Write Stuff Enterprises, Inc.
17. Claire Murphy, interview by Linda A. Firestone, digital recording, 16 May 2006, Write Stuff Enterprises, Inc.
18. Claire Murphy, interview by Linda A. Firestone, digital recording, 16 May 2006, Write Stuff Enterprises, Inc.
19. Claire Murphy, interview by Linda A. Firestone, digital recording, 16 May 2006, Write Stuff Enterprises, Inc.
20. Claire Murphy, interview by Linda A. Firestone, digital recording, 16 May 2006, Write Stuff Enterprises, Inc.
21. Claire Murphy, interview by Linda A. Firestone, digital recording, 16 May 2006, Write Stuff Enterprises, Inc.
22. Kevin Aubuchon, interview by Linda A. Firestone, digital recording, 4 May 2006, Write Stuff Enterprises, Inc.
23. Leon Silverstein, interview by Linda A. Firestone, digital recording, 4 May 2006, Write Stuff Enterprises, Inc.
24. Leon Silverstein, interview by Linda A. Firestone, digital recording, 4 May 2006, Write Stuff Enterprises, Inc.
25. Leon Silverstein, interview by Linda A. Firestone, digital recording, 4 May 2006, Write Stuff Enterprises, Inc.
26. Leon Silverstein, interview by Linda A. Firestone, digital recording, 17 January 2006, Write Stuff Enterprises, Inc.
27. Leon Silverstein, interview by Linda A. Firestone, digital recording, 17 January 2006, Write Stuff Enterprises, Inc.
28. Leon Silverstein, interview by Linda A. Firestone, digital recording, 17 January 2006, Write Stuff Enterprises, Inc.
29. Richard Jaffe, interview by Jeffrey L. Rodengen, digital recording, 19 October 2005, Write Stuff Enterprises, Inc.
30. Leon Silverstein, interview by Linda A. Firestone, digital recording, 27 April 2006, Write Stuff Enterprises, Inc.
31. Leon Silverstein, interview by Linda A. Firestone, digital recording, 27 April 2006, Write Stuff Enterprises, Inc.
32. Richard Jaffe, interview by Jeffrey L. Rodengen, digital recording, 19 October 2005, Write Stuff Enterprises, Inc.
33. Gerry Boylan, interview by Linda A. Firestone, digital recording, 12 May 2006, Write Stuff Enterprises, Inc.
34. Gerry Boylan, interview by Linda A. Firestone, digital recording, 12 May 2006, Write Stuff Enterprises, Inc.
35. Richard Jaffe, interview by Jeffrey L. Rodengen, digital recording, 19 October 2005, Write Stuff Enterprises, Inc.
36. Leon Silverstein, interview by Linda A. Firestone, digital recording, 4 May 2006, Write Stuff Enterprises, Inc.

37. Leon Silverstein, interview by Linda A. Firestone, digital recording, 4 May 2006, Write Stuff Enterprises, Inc.

Chapter Six Sidebar: Into the Fire

1. Don Hyde, interview by Linda A. Firestone, digital recording, 19 June 2006, Write Stuff Enterprises, Inc.

Chapter Six Sidebar: Planning for Success

1. Leon Silverstein, interview by Linda A. Firestone, digital recording, 4 May 2006, Write Stuff Enterprises, Inc.

Chapter Six Sidebar: Productivity on the Rise

1. Economic Profile and Trends, "Glass Industry Analysis," Office of Industrial Technology, Energy Information Administration, http://www.eia.doe.gov/emeu/mecs/ia98/glass/labor.html (accessed 12 March 2006).

Chapter Six Sidebar: Memory Lane

1. Cindy Stocklin, interview by Ann Gossy, digital recording, 4 April 2006, Write Stuff Enterprises, Inc.

Chapter Six Sidebar: One Job

1. Orpha Lopez, interview by Linda A. Firestone, digital recording, 16 May 2006, Write Stuff Enterprises, Inc.

Chapter Six Sidebar: What You Need to Know

1. Gerry Boylan, interview by Linda A. Firestone, digital recording, 12 May 2006, Write Stuff Enterprises, Inc.

Chapter Six Sidebar: Life's Work

1. Bridgio Lopez, interview by Linda A. Firestone, digital recording, 16 May 2006, Write Stuff Enterprises, Inc.

Chapter Seven

1. Leon Silverstein, interview by Linda A. Firestone, digital recording, 12 June 2006, Write Stuff Enterprises, Inc.
2. Leon Silverstein, interview by Linda A. Firestone, digital recording, 12 June 2006, Write Stuff Enterprises, Inc.
3. Leon Silverstein, interview by Linda A. Firestone, digital recording, 12 June 2006, Write Stuff Enterprises, Inc.
4. Leon Silverstein, interview by Linda A. Firestone, digital recording, 12 June 2006, Write Stuff Enterprises, Inc.
5. Leon Silverstein, interview by Linda A. Firestone, digital recording, 12 June 2006, Write Stuff Enterprises, Inc.
6. Joseph Devine, interview by Jeffrey L. Rodengen, digital recording, 2 December 2005, Write Stuff Enterprises, Inc.
7. Cliff Monroe, interview by Jeffrey L. Rodengen, digital recording, 19 October 2005, Write Stuff Enterprises, Inc.
8. Steven Perilstein, interview by Jeffrey L. Rodengen, digital recording, 2 December 2005, Write Stuff Enterprises, Inc.
9. Steven Perilstein, interview by Jeffrey L. Rodengen, digital recording, 2 December 2005, Write Stuff Enterprises, Inc.
10. Steven Perilstein, interview by Jeffrey L. Rodengen, digital recording, 2 December 2005, Write Stuff Enterprises, Inc.
11. Steven Perilstein, interview by Jeffrey L. Rodengen, digital recording, 2 December

2005, Write Stuff Enterprises, Inc.

12. Steven Perilstein, interview by Jeffrey L. Rodengen, digital recording, 2 December 2005, Write Stuff Enterprises, Inc.
13. Steven Perilstein, interview by Jeffrey L. Rodengen, digital recording, 2 December 2005, Write Stuff Enterprises, Inc.
14. Steven Perilstein, interview by Jeffrey L. Rodengen, digital recording, 2 December 2005, Write Stuff Enterprises, Inc.
15. Steven Perilstein, interview by Jeffrey L. Rodengen, digital recording, 2 December 2005, Write Stuff Enterprises, Inc.
16. Leon Silverstein, interview by Linda A. Firestone, digital recording, 12 June 2006, Write Stuff Enterprises, Inc.
17. Steven Perilstein, interview by Jeffrey L. Rodengen, digital recording, 2 December 2005, Write Stuff Enterprises, Inc.
18. Max Perilstein, interview by Jeffrey L. Rodengen, digital recording, 19 October 2006, Write Stuff Enterprises, Inc.
19. Max Perilstein, interview by Linda A. Firestone, digital recording, 9 February 2006, Write Stuff Enterprises, Inc.
20. Max Perilstein, interview by Linda A. Firestone, digital recording, 9 February 2006, Write Stuff Enterprises, Inc.
21. Max Perilstein, interview by Linda A. Firestone, digital recording, 9 February 2006, Write Stuff Enterprises, Inc.
22. Leon Silverstein, interview by Ann Gossy and Linda A. Firestone, digital recording, 17 January 2006, Write Stuff Enterprises, Inc.
23. Leon Silverstein, interview by Ann Gossy and Linda A. Firestone, digital recording, 17 January 2006, Write Stuff Enterprises, Inc.
24. Leon Silverstein, interview by Linda A. Firestone, digital recording, 12 June 2006, Write Stuff Enterprises, Inc.
25. Leon Silverstein, interview by Linda A. Firestone, digital recording, 12 June 2006, Write Stuff Enterprises, Inc.
26. Shawn Myers, interview by Jeffrey L. Rodengen, digital recording, 4 April 2006, Write Stuff Enterprises, Inc.
27. Don Hyde, interview by Linda A. Firestone, digital recording, 19 June 2006, Write Stuff Enterprises, Inc.
28. David Williams, interview by Linda A. Firestone, digital recording, 16 June 2006, Write Stuff Enterprises, Inc.
29. Jeffrey Ziesche, interview by Ann Gossy and Linda A. Firestone, digital recording, 31 March 2006, Write Stuff Enterprises, Inc.
30. Jeffrey Ziesche, interview by Ann Gossy and Linda A. Firestone, digital recording, 31 March 2006, Write Stuff Enterprises, Inc.
31. Jeffrey Ziesche, interview by Ann Gossy and Linda A. Firestone, digital recording, 31 March 2006, Write Stuff Enterprises, Inc.
32. Leon Silverstein, interview by Linda A. Firestone, digital recording, 12 June 2006, Write Stuff Enterprises, Inc.
33. Board meeting notes, 7 August 2002 and 11 February 2003.
34. Leon Silverstein, interview by Linda A. Firestone, digital recording, 12 June 2006, Write Stuff Enterprises, Inc.
35. Leon Silverstein, interview by Linda A. Firestone,

digital recording, 12 June 2006, Write Stuff Enterprises, Inc.
36. Leon Silverstein, interview by Linda A. Firestone, digital recording, 12 June 2006, Write Stuff Enterprises, Inc.
37. Donna Aubuchon and Ricky Silverstein, interview by Ann Gossy and Linda A. Firestone, digital recording, 31 January 2006, Write Stuff Enterprises, Inc.
38. Donna Aubuchon and Ricky Silverstein, interview by Ann Gossy and Linda A. Firestone, digital recording, 31 January 2006, Write Stuff Enterprises, Inc.
39. Daniel Wagner, interview by Linda A. Firestone, digital recording, 16 May 2006, Write Stuff Enterprises, Inc.
40. Daniel Wagner, interview by Linda A. Firestone, digital recording, 16 May 2006, Write Stuff Enterprises, Inc.
41. Daniel Wagner, interview by Linda A. Firestone, digital recording, 16 May 2006, Write Stuff Enterprises, Inc.
42. Leon Silverstein, interview by Linda A. Firestone, digital recording, 12 June 2006, Write Stuff Enterprises, Inc.
43. Leon Silverstein, interview by Linda A. Firestone, digital recording, 12 June 2006, Write Stuff Enterprises, Inc.
44. Mark Kuchenrither, interview by Ann Gossy and Linda A. Firestone, digital recording, 26 February 2006, Write Stuff Enterprises, Inc.
45. Mark Kuchenrither, interview by Ann Gossy and Linda A. Firestone, digital recording, 26 February 2006, Write Stuff Enterprises, Inc.
46. Max Perilstein, interview by Linda A. Firestone, digital recording, 9 February 2006, Write Stuff Enterprises, Inc.
47. Mark Kuchenrither, interview by Ann Gossy and Linda A. Firestone, digital recording, 26 February 2006, Write Stuff Enterprises, Inc.
48. Kevin McCarthy, interview by Linda A. Firestone, digital recording, 19 June 2006, Write Stuff Enterprises, Inc.
49. Richard Cravey and Kevin McCarthy, interview by Linda A. Firestone, digital recording, 19 June 2006, Write Stuff Enterprises, Inc.
50. Kevin McCarthy, interview by Linda A. Firestone, digital recording, 19 June 2006, Write Stuff Enterprises, Inc.
51. Richard Cravey and Kevin McCarthy, interview by Linda A. Firestone, digital recording, 19 June 2006, Write Stuff Enterprises, Inc.
52. Richard Cravey and Kevin McCarthy, interview by Linda A. Firestone, digital recording, 19 June 2006, Write Stuff Enterprises, Inc.
53. Mark Kuchenrither, interview by Ann Gossy and Linda A. Firestone, digital recording, 26 February 2006, Write Stuff Enterprises, Inc.
54. Mark Kuchenrither, interview by Ann Gossy and Linda A. Firestone, digital recording, 26 February 2006, Write Stuff Enterprises, Inc.
55. Leon Silverstein, interview by Linda A. Firestone, digital recording, 12 June 2006, Write Stuff Enterprises, Inc.
56. Leon Silverstein, interview by Linda A. Firestone, digital recording, 12 June 2006, Write Stuff Enterprises, Inc.
57. Leon Silverstein, interview by Linda A. Firestone, digital recording, 12 June 2006, Write Stuff Enterprises, Inc.
58. Leon Silverstein, interview by Linda A. Firestone,

digital recording, 12 June 2006, Write Stuff Enterprises, Inc.
59. Leon Silverstein, interview by Linda A. Firestone, digital recording, 12 June 2006, Write Stuff Enterprises, Inc.
60. Leon Silverstein, interview by Linda A. Firestone, digital recording, 12 June 2006, Write Stuff Enterprises, Inc.
61. Shirley Taton, interview by Linda A. Firestone, digital recording, 15 June 2006, Write Stuff Enterprises, Inc.
62. Donna Aubuchon and Ricky Silverstein, interview by Ann Gossy and Linda A. Firestone, digital recording, 31 January 2006, Write Stuff Enterprises, Inc.
63. Leon Silverstein, interview by Linda A. Firestone, digital recording, 12 June 2006, Write Stuff Enterprises, Inc.
64. Leon Silverstein, interview by Linda A. Firestone, digital recording, 12 June 2006, Write Stuff Enterprises, Inc.
65. Leon Silverstein, interview by Linda A. Firestone, digital recording, 12 June 2006, Write Stuff Enterprises, Inc.
66. Leon Silverstein, interview by Linda A. Firestone, digital recording, 12 June 2006, Write Stuff Enterprises, Inc.
67. Leon Silverstein, interview by Ann Gossy and Linda A. Firestone, digital recording, 17 January 2006, Write Stuff Enterprises, Inc.
68. Leon Silverstein, interview by Ann Gossy and Linda A. Firestone, digital recording, 17 January 2006, Write Stuff Enterprises, Inc.

Chapter Seven Sidebar: Customer Service Advantage

1. Gene Lomando, interview by Jeffrey L. Rodengen, digital recording, 18 October 2005, Write Stuff Enterprises, Inc.

Chapter Seven Sidebar: More to Learn

1. Dan Haney, interview by Linda A. Firestone, digital recording, 26 June 2006, Write Stuff Enterprises, Inc.

Chapter Seven Sidebar: Thug for a Day

1. Mark Kuchenrither, interview by Linda A. Firestone, digital recording, 5 May 2006, Write Stuff Enterprises, Inc.

Chapter Seven Sidebar: Sister Assistance

1. Natalie Reed, interview by Ann Gossy, digital recording, 12 June 2006, Write Stuff Enterprises, Inc.

Chapter Seven Sidebar: Memory Lane

1. Max Perilstein, interview by Linda A. Firestone, digital recording, 9 February 2006, Write Stuff Enterprises, Inc.

Chapter Seven Sidebar: The Sydney Awards

1. Arch newsletter.

Chapter Seven Sidebar: The Moral of the Story …

1. Mark Kuchenrither, interview by Linda A. Firestone, digital recording, 5 May 2006, Write Stuff Enterprises, Inc.

Chapter Seven Sidebar: A Company with Heart

1. Rose Oliver, interview by Linda A. Firestone, digital recording, 16 May 2006, Write Stuff Enterprises, Inc.

Appendix

Starrett City

1. Leon and Ricky Silverstein, interview by Linda A. Firestone, digital recording, 28 June 2006, Write Stuff Enterprises, Inc.
2. Leon and Ricky Silverstein, interview by Linda A. Firestone, digital recording, 28 June 2006, Write Stuff Enterprises, Inc.
3. Leon and Ricky Silverstein, interview by Linda A. Firestone, digital recording, 28 June 2006, Write Stuff Enterprises, Inc.
4. Leon and Ricky Silverstein, interview by Linda A. Firestone, digital recording, 28 June 2006, Write Stuff Enterprises, Inc.
5. Leon and Ricky Silverstein, interview by Linda A. Firestone, digital recording, 28 June 2006, Write Stuff Enterprises, Inc.

Time Warner Building

1. Leon and Ricky Silverstein, interview by Linda A. Firestone, digital recording, 28 June 2006, Write Stuff Enterprises, Inc.
2. Leon and Ricky Silverstein, interview by Linda A. Firestone, digital recording, 28 June 2006, Write Stuff Enterprises, Inc.
3. Leon and Ricky Silverstein, interview by Linda A. Firestone, digital recording, 28 June 2006, Write Stuff Enterprises, Inc.
4. Leon and Ricky Silverstein, interview by Linda A. Firestone, digital recording, 28 June 2006, Write Stuff Enterprises, Inc.
5. Leon and Ricky Silverstein, interview by Linda A. Firestone, digital recording, 28 June 2006, Write Stuff Enterprises, Inc.

Farhat Aluminum

1. Leon Silverstein, interview by Linda A. Firestone, digital recording, 25 August 2006, Write Stuff Enterprises, Inc.
2. Leon Silverstein, interview by Linda A. Firestone, digital recording, 25 August 2006, Write Stuff Enterprises, Inc.

The Atlantis Hotel

1. Leon and Ricky Silverstein, interview by Linda A. Firestone, digital recording, 28 June 2006, Write Stuff Enterprises, Inc.

Anheuser-Busch Brewing Company

1. Leon and Ricky Silverstein, interview by Linda A. Firestone, digital recording, 28 June 2006, Write Stuff Enterprises, Inc.
2. Leon and Ricky Silverstein, interview by Linda A. Firestone, digital recording, 28 June 2006, Write Stuff Enterprises, Inc.

Universal Studios

1. Leon and Ricky Silverstein, interview by Linda A. Firestone, digital recording, 28 June 2006, Write Stuff Enterprises, Inc.

Entourage Martini Bar

1. Jeffrey Ziesche, interview by Linda A. Firestone, digital recording, 3 August 2006, Write Stuff Enterprises, Inc.

The Gato Cigar Factory

1. Jeffrey Ziesche, interview by Linda A. Firestone, digital recording, 3 August 2006, Write Stuff Enterprises, Inc.
2. Jeffrey Ziesche, interview by Linda A. Firestone, digital recording, 3 August 2006, Write Stuff Enterprises, Inc.

INDEX

Page numbers in italics indicate photographs.

A

Abu Dhabi (United Arab Emirates), 116–17
Adams, Jim, 111
Adams Glass Company, 26, 29
advertising, by Arch Aluminum, *88*, 96–97, *99*
Akron (Ohio), 80, 94
Aldora Aluminum Products (firm), 29–33, *29*
  bought by Chromalloy, 35
  under Chromalloy, 38–40
  moves to larger facility, 30–31, *31–32*
  trade show booth of, *35*
Alpern, Bob, 44, *44*, *45*
  bought out by Bobby Silverstein and Marvin Cravetz, 52
aluminum
  Arch Aluminum products, 101
  Architectural Glass Products sales of, 28–29
  beginnings of industry for, 12
  extruding, 102–3, *102*
  painting of, 60–61
  technology of production of (in 1950s), 23
Aluminum Association, 12
Aluminum Shapes (firm), 71, 73
  Arch Aluminum divides from, 81–83
  Arch Aluminum merges with, 75–77
Amarlite (firm), 26, 28, 29
  purchased by Arch Aluminum, 71–73, *72*
American Architectural Products (firm), 100
American Flat Glass (AFG), 111
American Float Glass (firm), 43
American Glassmith (firm), 100
American Institute of Architects (AIA), 58
American Safety Glass (firm), 29
Anderson, Donny, viii, 115–16, *116*
Anheuser-Busch brewery, 118
Arch Aluminum & Glass Company, Inc.
  acquisitions by, 80, 101–2, 112–13
  advertising by, *88*, 96–97, *99*
  aluminum products of, 101
  awards for, 101
  beginning in business of, 45–47
  Binnings Pan American and, 102–3
  catalogs of, *74*
  computerization of, 53–55, 79–81
  Coolidge purchased by, 89–90
  customers of, 59–60
  customized software developed for, 61
  Daniel Wagner hired by, 103–4, *104*
  in dispute with Marvin Cravetz and Steven Marcus, 83–86
  divides from Aluminum Shapes, 81–83
  employees of, 57–59
  employees of, health care options for, 107–8
  expansion of (1990s), 70–73
  expansion of (2004–2005), 110–12
  first headquarters of, *46*
  founded by Bobby Silverstein, 11, 41
  future of, 113
  growth of and acquisitions by, 49–52
  initial investments in, 44–45
  insulated glass by, *70*

internship program of, 109–10, *110*
logo for, *42*
Long Point Capital invests in, 86–87
Mark Kuchenrither CFO of, 104–7
after merger with Aluminum Shapes, 75–78
mirror production by, 66–67
new facilities built for, 78–79
origins of name of, 45
Orlando facility of, 52–53
oversized sheets of glass manufactured by, 68–69, *69*
painting of aluminum products of, 60–61
recapitalization of, 108–9
signature projects of, 114–21
Leon Silverstein at, 47–49
suppliers for, 43–44
Sydney Awards to employees of, 105, *105*
Tamarac facility of, *77*
tempered glass produced by, 65–66
United Glass Corporation and, 91–94
Arch Aluminum of Pittsburgh (firm), 91–94, 101
Arch America (firm), 75–77
Architectural Glass Products, Inc., *22*, 23–28, *25*
Aldora Aluminum and, 29–30, *29*
bought by Chromalloy, 35
under Chromalloy, 39, 40
customers of and suppliers to, 30
moves to larger facility, 30–31, *31–32*
architecture
LEED criteria for, 58
modernism in, 11, 12
signature projects, 114–21
use of glass in, 57
to withstand hurricanes, 63–65, 75, *92*
Arch Mirror West (firm), 99
Arch Texas (firm), 52
Ashley, Gerald, viii, 57–58, *105*
Atlanta (Georgia), 49, 51–52, 78
Atlantis Hotel (Bahamas), 117–18, *117*
Aubuchon, Donna Silverstein, 25–26, 36, 54, *79*
at Apple, 53
bat mitzvah of, *38*
in college, 41
at Leon Silverstein's bar mitzvah, *39*
on Ricky Silverstein, 103
in Tamarac, 107
working for Arch Aluminum, 111–12
Aubuchon, Kevin, viii, *53*, *91*
computerization of Arch Aluminum under, 53–55, *54*, 61, 79–81, *79*
in Tamarac, *107*
on Y2K bug, 87
Aubuchon, Maddie, 109, *110*
Aubuchon, Rachael, 109, *110*
auctions, 64–66
Audette, Jackie, viii, *108*
Austell (Georgia), *50*, 52, 55, *64*, 78

B

Bahamas, 117–18, *117*
Baskwell, Paul, viii
Battendorf (Iowa), 67
Bay Mirror (firm), 99
Beaver, Don, viii, 39, 49, 52, *64*
Beerman, Samantha, viii
Binnings Pan American (BPA), 102–3
Bonnell, William L., 29
boom trucks, *38*
Boston (Massachusetts), 41
Boylan, Gerry, viii, 86
Bridgeport (New Jersey), 97
Britt, Billy, viii
Brock, Ron, viii, 62, 65, 66
Buchan, Ralph, *27*
buildings, *see* architecture

C

California, 97–99, 104
Carolina Mirrors (firm), 27, 29, 37, 43
Chamberlain, Wilt, 17
charities, 109
Chromalloy (firm)
Arch Aluminum and, 44
Architectural Glass Products and Aldora Aluminum Products bought by, 35
Bobby Silverstein at, *34*, 36–41, 62
trade show booth of, *35*
Claros, Carlos, *96*
Coben, Paula, viii
Coleman, Cathy, viii
Collette, Roger, viii, 24, *24*, 59
at Aldora plant, 48, *48*
Collins Glass (firm), 29
Columbus (Ohio), 112
Congress Bank, 71
Coolidge (firm), 89–90
Copeland, Paul, 38, 39, 49, *49*
Cove, Douglas, viii, 45, 48, 54
Cravetz, Marvin, *44*
Alpern bought out by, 52
Aluminum Shapes and, 71
as Bobby Silverstein's childhood friend, 17
division of Arch Aluminum and Aluminum Shapes financed by, 82
investment in Arch Aluminum by, 44, 83–86
Cravey, Dick, 108, 109
Cravey, Green & Wahlen (CGW), 108

Cystic Fibrosis Association, 109

D

Dallas (Texas), 39, 99–100, 112
D & S Glass (firm), 99–100
Davis, Richard, viii, 51
Denver (Colorado), 111
Devine, Joseph, viii, 90
Dudoff, Harold, 26, 116–17
Dunn, John, 43, 44, *44*, 46

E

energy issues, 57, 58
Entourage martini bar, 118, *118*
environmental issues, *57*, 58
extruding aluminum, 102–3, *102*
Extrudite Metal Works (firm), 14

F

Farhat Aluminum Company, 116–17, *116*
float glass, 20, 57
Florida
- Architectural Glass Products in, 23–28
- Hurricane Andrew in, 63, 64
- hurricanes in, 71, 97, 121
- mirrors manufactured in, 62
- Syndey Silverstein buys Pritchard Paint and Glass in, 21
- Silverstein family moves back to, 48

Fort Lauderdale (Florida), 49, 52
Fort Myers (Florida), 49–50, 71
Fort Pierce (Florida), 62, 63, 66

G

Garcia, Gabe, viii, 58, *58*
Gato, Eduardo, 120
Gato Cigar factory (Key West, Florida), 120–21, *120*
GE Capital (firm), 90
George, Les, viii
Glass Alum (firm), 115, 116
glass and glass industry
- architectural uses of, 57
- assembly-line manufacture of, 11–12
- boom trucks for, *38*
- classical manufacture of, *13*, *14*
- consolidation in (1990s), 70–73
- environmental concerns and, 57, 58
- for green buildings, 58
- history and types of, 20
- insulated glass, *70*
- laminated glass, *89*
- in mirrors, 62–64
- in modernist architecture, 11
- oversized sheets of, 68–69, *68–69*, 115, 118
- productivity in, 81
- standards for, 91
- technology of production of (in 1950s), 23
- tempered glass, 65–66

Glass Tex (firm), 111
Gloucester City (New Jersey), 31–33
Goldenberg Candy Company, 47
Goodrich, Garth, viii, 61
Great Depression, 11
green buildings, 58
greenfield facilities, 78, 83, 94

H

H. Perilstein Company, 12–14, *13*, 17
- purchased by Chromalloy, 35

Hall, Charles Martin, 12
Hamilton Glass Company, 20–21
Handi, Mary, viii, 62, *62*, 63, 99
Handi, Numa, viii, 62–63, *62*, 66, 99
Haney, Dan, 70, 93, 100
Hankerson, Fred, viii
Harding (firm), 73
Hardtemp (firm), 73
Hayes, Ed, 62
health care, for Arch Aluminum employees, 107–8
Hires Turner (firm), 35
Horses and the Handicapped of South Florida, Inc., 109
Houston (Texas), 111
hurricanes, 21, 71
- Andrew, 63, 64
- Arch Aluminum materials for, *92*, 97, 121
- building standards for, 64–65, 75

Hyde, Don, viii, 76, 100

I

impact glass, 64
Indianapolis (Indiana), 70, 80
insulated glass, *70*
International Brotherhood of Teamsters (IBT), 76
International Union of Architects (IUA), 58
internship program, 109–10, *110*
Iowa, 66, 79

J

Jackson, Art, viii, 59
Jackson, Harold, 82
Jackson (Mississippi), 80, 82–83
Jaffe, Richard, viii, 75, 84, 86
Jaffray, Piper, 108
Jesse (dog), *108*

K

Kansas City (Missouri), 73
Kendall, Steve, 71, 73
- on division of Arch Aluminum and Aluminum Shapes, 82

after merger, 75–77
Key West (Florida), 120–21, *120*
King of Prussia (Pennsylvania), 36–37
Korff, Max, 41
KraftMaid Cabinets (firm), 112
Kramer, Arnold, viii, 41
Kuchenrither, Mark, viii, 95, *95*, 112
as CFO, 104–7
internship program managed by, 109–10

L

labor unions, 25, 76, 83, 114
laminated glass, 20, 57, *89*
Lawson, Tony, viii, 97
Leadership in Energy and Environmental Design (LEED) Green Building Rating System, 58
Lerner, Gale, 86
Lomando, Gene, viii, 91
London, Ray, viii, 52, 83
Arch Aluminum of Dallas purchased by, 72
Arch Aluminum of Dallas sold by, 112
as manager of Arch Aluminum of Dallas, 39
Long Point Capital (firm), 86–87, 104
Lopez, Bridgio, viii, 87
Lopez, Orpha, viii, 84

M

Marcus, Steven, 82–86
MASCO, 104
Max (dog), *61*
Mayer, Lothar, 86
McCalla, Jeffrey, *94*
McCarthy, Kevin, 108, 109
McKay, Gene, 111
Medley (Florida), 30–31, *31*
Memphis (Tennessee), 70, 80, 82, 83, 87
Miami (Florida), 41
expansion of facility in, 112–13
Miami-Dade County (Florida), 64–65, 75
mirrors, 62–64, 97–99
modernism (in architecture), 11, 12
Monroe, Cliff, viii, 90–91
Moses, Alan, viii, 44–45
Murphy, Claire, viii, *79*, 80–81
Murphy, Neil, viii, 59
Myers, Shawn, viii, 100

N

National Glass Association, *84*
New York
Starrett City, 114–15, *114*
Time Warner Building, 115–16, *115*
Norcross (Georgia), 78

O

Orlando (Florida), 52–53, 65, 118

P

Palie, Margie, 17, 26, *27*
Parcell, Paul, viii, 23, 60, 61
Pennsauken (New Jersey), 33, 40, 52
Perilstein, Beth, 100
Perilstein, Dora, *see* Silverstein, Dora Perilstein
Perilstein, Fagie Mollie, *10*
Perilstein, Harris, *10*, 12, 13, 91, *91*
Perilstein, Josh, *93*
Perilstein, Louis, 14, 15
Perilstein, Marcie, *91*
Perilstein, Max, viii, 15, *91*, 94–97, 100
on Mark Kuchenrither, 105
Perilstein, Steven, viii, 91–94, *91*, *93*, 96
Perilstein Distributing Corporation (PDC), 91, 100
Perilstein family, 12–13, *16*, 17–20
Silverstein family and, 94–96
United Glass Corporation of, 91–94
Pittsburgh (Pennsylvania), 94
Pompano Beach (Florida), 52, 65, 77
Portland (Oregon), 80, 102
Powers, Don, viii, 79–80
Pritchard Paint and Glass (firm), 21, 23, *24*

R

Rabinowitz, Abe, 26
Reed, Natalie Silverstein, viii, *16*
birth of, 14
working for Arch Aluminum, *64*, 87, 97
Reifman, Paul, viii, 15
Reifman, Ruth, viii, 15
Resister Impact line of hurricane materials, *92*, 97
Reuben Glass (firm), 35
Rodriguez, Mike, 102–3, 112
Rogers, Mike, viii, 38, 71
Rothenberg, David, *110*

S

safety glass, 20–21
St. Louis (Missouri), 118
Salt Lake City (Utah), 112
Sarasota (Florida), 50
sheet glass, 20
Sheets, Roger, 48
signature projects of Arch Aluminum, 114–21

Silverstein, Beani, viii, *27*, 28, 33, *103*
at Austell, *64*
becomes a mother, 25
on Bobby Silverstein leaving Chromalloy, 40–41
death of father of, 17
at Donna Silverstein's bat mitzvah, *38*
on family business, 13
honeymoon of, 21
at Leon Silverstein's bar mitzvah, *39*
logo for Arch Aluminum by, *42*, 45
moves to new home, 26
on move to Philadelphia, 36–37
on Syndey Silverstein, 23
Silverstein, Cindy, 36, 41
on Bobby Silverstein's love of cars, 82
at Donna Silverstein's bat mitzvah, *38*
at Leon Silverstein's bar mitzvah, *39*
Silverstein, Donna, *see* Aubuchon, Donna Silverstein
Silverstein, Dora Perilstein, 13–16, *15*, *16*
Silverstein, Leon, viii, *91*, *108*, *111*
on acquisitions, 80
at Arch Aluminum, 20, 57, *57*, 60–61
Kevin Aubuchon and, 55
at auctions, 64–66
bar mitzvah of, *39*
Binnings Pan American and, 102–3
birth of, 26
in California, 97–99
childhood of, 33, 36
on Coolidge purchase, 89–90
in dispute with Marvin Cravetz and Steve Marcus, 84
on division of Arch Aluminum and Aluminum Shapes, 82
at Donna Silverstein's bat mitzvah, *38*
expansion of Arch Aluminum under, 70–73
on expansion of Miami facility, 112–13
on expansion of Tamarac facility, 111
on future of Arch Aluminum, 113
on Jackson facility, 83
Jeff Ziesche and, 101
Long Point Capital and, 86–87
Mark Kuchenrither and, 104–5
after merger with Aluminum Shapes, 75–77
Orlando facility and, 52–53
Steven Perilstein and, 92–94
on signature projects, 114–18
starts work at Arch Aluminum, 47–48
Sunshine Mirror and, *56*, 62–63, 66, 67
Sydney Award given by, *105*
in Tamarac, 78, *107*
working for Arnold Kramer, 41
on Youngsville facility, 78–79
Silverstein, Marnie, 109, *109*, 110
Silverstein, Natalie, *see* Reed, Natalie Silverstein
Silverstein, Paula (Helen), 14, *16*, 60
Silverstein, Ricky, viii, 41, *91*
birth of, 36
as chief of operations, 87, 103
at Donna Silverstein's bat mitzvah, *38*
on glass for Time Warner building, 115
starts working for Arch Aluminum, 61
Sydney Award given by, *105*
in Tamarac, *107*
Silverstein, Robert Harris (Bobby), viii, *16*, *27*, *32*, *43*, *91*, *103*, *105*, *111*
Aldora Aluminum and, 29–30, 33
Arch Aluminum founded by, 11, 41, 43–44
Arch Aluminum's beginnings under, 45–47
at Architectural Glass Products, 23–28
Aubuchon and, 53–55
at Austell, *64*
birth of, 14
cars owned by, 82
childhood of, 16
at Chromalloy, *34*, 36–40
on customers of Arch Aluminum, 59–60, *60*
in dispute with Marvin Cravetz and Steve Marcus, 84
on division of Arch Aluminum and Aluminum Shapes, 82
at Donna Silverstein's bat mitzvah, *38*
expansion of Arch Aluminum under, 70–73
on family business, 13
high school days of, 17
high school graduation photo of, *17*

honeymoon of, 21
Jeff Ziesche and, 101
leaves Chromalloy, 40–41
at Leon Silverstein's bar mitzvah, *39*
Long Point Capital and, 86–87
Mark Kuchenrither and, 104–6
MG sports car bought by, *26*
Orlando facility and, 52–53
Natalie Silverstein Reed and, 97
on Sydney Silverstein, 15
Sunshine Mirror and, 62–63, 66
in Tamarac, *107*
at West Penn Mirror, 20
Silverstein, Sydney, *11*, 13, *15*, *16*, *32*
Aldora Aluminum under, 29–31
Architectural Glass Products and Aldora Aluminum Products sold to Chromalloy by, 35
Architectural Glass Products under, 23–28, *24*, *25*, *27*
death of, 36
family life of, 16–17
health of, 15
marries Dora Perilstein, 14
in new Medley facility, *31*
Natalie Silverstein Reed on, 97
West Penn Mirror and, *18–19*, 20–21
Silverstein family, 14–17, *107*
Aluminum Shapes and, 77, 82
Cliff Monroe on, 90
Donna Silverstein's bat mitzvah, *38*
Gerry Boylan on, 86
Leon Silverstein's bar mitzvah, *39*
new home for, 36, *36–37*
Perilstein family and, 94–96
real estate investments of, 67
recapitalization of investments of, 108–9
Solomon Glass (firm), 41
Star, Ira, 86
Starrett City (Brooklyn, New York), 114–15, *115*
Stern, Scott, *110*
Stocklin, Cindy, viii
Studenski, Arnie, viii, 33, 87
at Aldora Aluminum, 29
at Arch Aluminum, 46, 47, *47*
at Architectural Glass Products, 23
in United Arab Emirates, *116*, 117
Sunshine Laminating (firm), 63, 64
Sunshine Mirror (firm), *56*, 62–64, 66–67
Superior Shower Door Glass Line, 97
"swat team," 100
Sydney Awards, 105, *105*

T

Tamarac (Florida), *77*, 78, 83
expansion of facility at, 110–12
Silverstein family in, *107*
Taton, Gene, 111
Taton, Shirley, 111
Tech-a-lite (firm), 49, *49*, *50*, 52
tempered glass, 20, 57, 65–66
Texas, 99–100, 111
Thermo Glass Products (firm), 39, 49, 50, *52*
Time Warner Building (New York), 115–16, *115*
Toby (dog), 54, *54*
Trico, John, 117
Trident (California), 80
Trident Industries (firm), 104
Tropical Glass (firm), 29, 91

U

Udenfriend, Marty, 31–33
United Arab Emirates, 116–17
United Glass Corporation (UGC), 91–94
U.S. Green Building Council, 58
Universal Studios (Orlando, Florida), 118

V

Vartanian, Tanya, *110*
Vega Glass (firm), 39
Villa Rica (Georgia), 78, *78*, 79, *79*, 83

W

Wagner, Daniel, viii, 103–4, *104*
Wareham, Jeff, 90
Weber, Terri, 80
Wells, Greg, 94
West Penn Mirror, Inc., 16, *18–19*, 20–21
Westshore Glass (firm), 62–63, 66–67
William L. Bonnell Company, 29, 43, 44, *44*
Williams, David, viii, 100, 112
Wind-Tech Products, Inc., 120–21

Y

Youngsville (North Carolina), 78–79, *79*

Z

Zabric, Natalie, 54, *54*
Ziesche, Jeff, viii, *99*, 101, 118, 121

Arch Aluminum & Glass Co., Inc.

YOUR TRUE SINGLE SOURCE®